AF735713

the mechanical properties of matter

CRANFIELD INSTITUTE
OF TECHNOLOGY
MAIN LIBRARY
ACC NO 042192
CLASS 539.3

THE BRIDGE SERIES

editor

F. C. BROWN M.A., B.Sc., F.R.I.C.
Senior Lecturer, Institute of Education, University of London

energy and bonding

MICHAEL HUDSON Ph.D., B.Sc., A.R.I.C.

M. T. SPRACKLING Ph.D., A.Inst.P.
Lecturer in Physics, Queen Elizabeth College
University of London

the mechanical properties of matter

THE ENGLISH UNIVERSITIES PRESS LTD
St Paul's House Warwick Lane London EC4

First Printed 1970

Copyright © 1970
M. T. SPRACKLING
All rights reserved. No part of this publication may be reproduced or transmitted in any form or by any means, electronic or mechanical, including photocopy, recording or any information storage and retrieval system, without permission in writing from the publisher.

SBN 340 05180 9
Printed and bound in Great Britain for The English Universities Press Ltd by W & J Mackay & Co Ltd, Chatham

editor's foreword

The transition from school to university or college is, for many students a difficult process and this is reflected in the relatively high proportion of failures occurring in the first years of university courses. Contributory causes include a significant increase in pace, drastic changes in teaching and learning patterns and an increased emphasis on private study and book-usage.

The purpose of this new series of texts is to bridge the gap between school and university or college by supplying concise and up-to-date presentations of important topics in the broad spectrum of science, mathematics and engineering. Some of the topics will be ones less frequently encountered in the sixth form curriculum, while others will extend and enrich more familiar themes. The books will concentrate on essential principles, with worked-out problems where appropriate.

The texts should provide valuable stimulus to the able sixth former, prior to leaving school, and they should also be of the greatest assistance to the first year student in university, college of education or college of further education.

F. C. Brown

acknowledgements

In the course of writing this volume, I have drawn on a number of books and articles, and in particular I should like to acknowledge my debt to the article 'Large-Scale Properties of Matter', by D. Tabor, which appeared in Volume 6 of *Contemporary Physics*, and the book *Properties of Matter*, by H. N. V. Temperley.

I should like to thank Dr M. McInally, Dr G. R. Field and Professor R. E. Burge for reading and commenting on various chapters of the book, and Mrs J. Bignall and Mrs R. Wilman for typing the manuscript.

M.T.S.

CRANFIELD INSTITUTE OF TECHNOLOGY
LIBRARY

preface

The aim of this book is to give a survey of the mechanical properties of matter at a level suitable for a first-year university course. I have endeavoured to treat the various phenomena included under the above heading both in terms of large-scale concepts and in terms of the behaviour of the basic particles of which matter is composed. Considerable emphasis has been placed on the atomic or molecular nature of matter and the way in which the response of the bulk material to applied forces can be explained in terms of interparticle forces and particle kinetic energy. This approach gives a unity to the various topics treated that is not always apparent.

Examples have been placed at the end of some chapters. Many of these examples serve not only to give the reader an opportunity to test his understanding of a topic, but also to amplify and illustrate the text.

October 1969 M.T.S.

contents

Chapter 4. Fracture

Chapter 5. Liquid Viscosity

Chapter 6. The Viscosity of Gases

Chapter 7. Surface Effects

1
solids, liquids and gases

1.1 Introduction

One way of classifying substances is according to their state of aggregation or 'phase'. Three basic phases are usually recognized, namely, solids, liquids and gases. A solid is an aggregation of matter that occupies a definite volume at a given temperature, and has a definite shape, while a liquid has a given volume at a given temperature, but its shape is determined by that of the containing vessel, except in the case of very small drops. A gas fills completely all the space made available to it, and necessarily takes the shape of the boundaries of the space. The term phase is also used to describe different modifications of the arrangement of the basic particles constituting a given solid.

All materials are gaseous at sufficiently high temperatures and all condense into liquids as the temperature is lowered. In addition, all, except helium, form solids under their own vapour pressure as the temperature is lowered still further. The basic particles of which matter is composed exert forces on each other which, in the condensed phases, are able to bind the particles together and are thus known as 'binding forces'. When the basic particles become linked in this way 'chemical bonds' are said to be formed between them. Each particle is able to form a small, and usually fixed, number of bonds.

The actual laws of force between the basic particles of matter depend on the electronic structure of the particles, giving rise to different types of bond, some of which are strongly directional and others which are not. Examples of different types of bond are ionic, covalent, and van der Waals bonds. The ionic bond results when electrons are transferred from atoms of one type to those of a second type, so that the material is composed of positive and negative ions. In the covalent bond, valence electrons are shared between two atoms to an appreciable extent, giving a bond with marked directional properties. The van der Waals bond is a weak electrostatic interaction which occurs even in atoms and molecules showing no permanent dipole moment. It arises from the fluctuating dipole moment associated with the instantaneous positions of the electrons in the atom. In many materials one type of bond predominates so that, for example, a certain solid may be referred to as an ionic solid. In other cases the type of bonding is not so clear cut. The basic particles of which matter is composed may be atoms, ions or chemical molecules. Among the commonly occurring solids, examples of all three types of particle are found. Rocksalt, for example, contains positive sodium ions and negative chlorine ions. Diamond is composed of carbon atoms held together by covalent bonds, and in paraffin wax there are individual molecules held together by van der Waals forces. In metals the atoms are, in fact, positive ions in a sea of electrons. In view of this diversity, the term particle will, in general, be used to describe the basic constituents of solids, whatever their nature. On the other hand, for a large number of gases and liquids the basic constituent is the molecule and this term will usually be used in these cases.

Despite the differences in the nature of the binding forces between particles, most materials show similar properties of aggregation as the temperature is lowered, and similar responses to applied forces. Hence the form of the interaction between particles must be qualitatively the same in all cases. One aim of this book is to establish the qualitative form of the interparticle force law, and to use it to discuss the mechanical properties of matter, that is, the response of matter to applied forces.

The different phases arise from the competition between the interparticle forces, which try to produce a regular arrangement of the particles, and the particle kinetic energy,* which tries to destroy such an arrangement. Applied pressures can affect the role played by the interparticle forces, and thus the phase in which any particular substance is found depends on its temperature and on the external pressure.

For many substances the change from one phase to another as, say, the temperature is changed at constant pressure, is not very distinct, but for simple pure substances consisting of one type of molecule a clear distinction can be made between the solid, liquid and gaseous phases. Consider a fixed mass of such a substance contained in a cylinder by a

* The sum of the particle kinetic energy and the potential energy arising from the particle interactions is called the 'internal energy' of the system under consideration. The internal energy can be changed by doing work on the system or by supplying heat to it, the term heat being reserved for a flow of energy taking place under a temperature gradient. Clearly, both the kinetic and potential energies of the particles are temperature dependent.

piston. When the substance is in any equilibrium state the state can be specified in terms of properties that refer to the complete mass, e.g. temperature and pressure. These properties are referred to as thermodynamic co-ordinates and by their nature can refer only to equilibrium states, in which they have the same value throughout the whole mass of the substance. Let pressure p, volume V and temperature T be chosen as the co-ordinates for the particular mass of substance considered, and let the equilibrium values of V corresponding to different values of p at

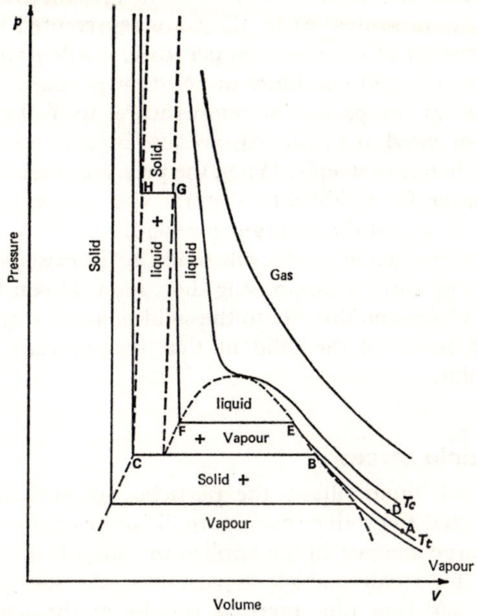

Figure 1.1

a fixed temperature be plotted on a p-V graph. The succession of points obtained can be considered to lie on a smooth curve. Such a curve is called an isotherm. If the above experiment is repeated at different temperatures and over a wide range of pressures, the results obtained are as shown in fig. 1.1. Here it is assumed that the substance expands on melting.

For all temperatures above T_c, known as the 'critical temperature', the substance remains in the gaseous phase whatever the value of the applied pressure. At temperatures T_c and below it is possible to liquify the substance simply by increasing the pressure. To distinguish between these two possibilities in the gaseous phase, it is customary to refer to 'gases' when the temperature is above T_c and 'vapours' when the temperature is T_c or below. At a temperature T_t, which is below T_c, the substance exists as a vapour for low values of p, as in the state represented by A in fig. 1.1, but as the pressure is increased a state B is reached in which solid, liquid and vapour coexist in equilibrium. T_t is known as the 'triple point temperature'. Further compression of the substance at T_t takes place at constant pressure until the whole of the substance is in the solid phase at C. Solids are almost incompressible and hence a large increase in pressure is needed to produce a relatively small further decrease in volume.

In the temperature range between T_c and T_t the substance exists as a vapour at low pressures, as in the state represented by D. As the volume is decreased at constant temperature, condensation of vapour to liquid starts at E and continues at constant pressure (the saturated vapour pressure at the particular temperature) until the whole of the vapour has condensed at F. Liquids are also almost incompressible, so that the curve then rises steeply. When the pressure reaches a sufficiently high value, as at G, solidification starts and continues at constant pressure until, at H, all the substance is solid.

At temperatures below T_t the substance again exists as a vapour at sufficiently low pressure. Compressing the vapour at constant temperature causes condensation directly to the solid phase at constant pressure (the vapour pressure of the solid at that temperature) until all the substance is solid.

1.2 Interparticle forces

In the solid and liquid phases the particles are very close together. Fig. 1.1 shows that to produce even a small compression of these phases needs a very large increase in the applied pressure. From this it may be inferred that, for small particle separations, the interparticle forces are repulsive, and that they increase rapidly as the separation is reduced.

Solids and, under certain conditions, liquids can withstand applied tension which tries to separate the particles, so that for larger particle separations the interparticle forces are attractive. The attractive force must become stronger with small increases in the particle separation, since the force needed to extend a solid increases with the extension, at least for a small range of increase. However, the attractive forces must be short-range forces, that is, they fall off rapidly with distance, since gases obey Boyle's Law reasonably well under conditions in which the density is not very different from that of the liquid, yet Boyle's Law can be deduced using the kinetic theory of gases only if the attractive interparticle force is zero. Thus, the variation with interparticle distance x of the interparticle force F for a particle and a single neighbour must take the general form shown in fig. 1.2a. The force is repulsive for small interparticle distances, becoming stronger with decrease of this distance, and is attractive at larger distances, reaching a maximum and falling practically to zero very rapidly. At some value x_o of the interparticle distance the interparticle force is zero. This is the equilibrium separation for the two particles, provided that neither has any kinetic energy. The equilibrium is stable, since, if the value of x is reduced the particles repel each other, and if it is increased they attract. The potential energy of a pair of particles is measured by the work that must be done against interparticle forces to produce a

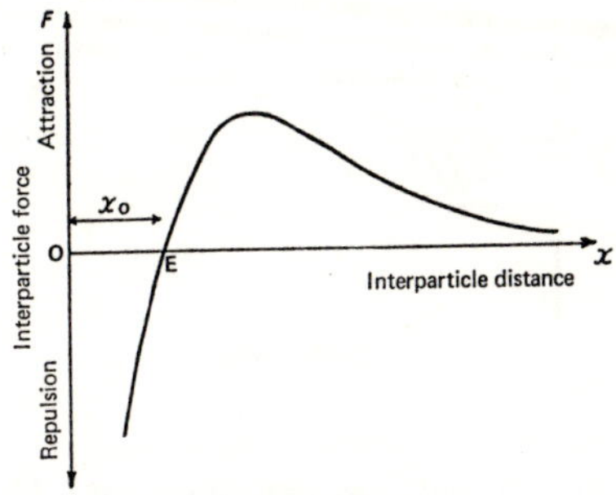

Figure 1.2a

particular configuration, starting with an infinite interparticle distance. The change in potential energy dU of a pair of particles resulting from a change in their separation of dx is $d\,U = -F\,dx$. The minus sign ensures that spontaneous processes result in a reduction of potential energy. Thus, fig. 1.2a can be represented equally well as a distribution of potential energy and interparticle distance, as in fig. 1.2b. The equilibrium separation x_o here corresponds to the minimum of the U–x curve, provided the particles do not possess any kinetic energy. The form of the U–x curve is for many purposes sufficiently well represented by the following empirical equation

$$U = -\frac{A}{x^{\mathrm{m}}} + \frac{B}{x^{\mathrm{n}}}.$$

A and B are positive constants and n must be greater than m, so that the positive term is dominant as x becomes very small, but falls to zero more rapidly than the negative term when x is larger.

In any real aggregation of particles, any given particle will experience forces due to its nearest neighbours and also those that are farther away. However, since the interparticle attractive forces are of very short range, the interactions with nearest neighbours will be most important, so it is reasonable to expect that the potential energy-interparticle distance curve for particles with many neighbours will take the general form shown in fig. 1.2b, though the exact shape of the curve, and, in particular, the value of x_o and of ΔU, will be slightly changed. Thus, if they do not possess kinetic energy, the particles will take up positions relative to their neighbours with the interparticle distance approximately x_0. Arranging particles so that all nearest neighbours have the same separation gives rise to a regular three-dimensional array. A three-dimensional array, or lattice, is typical of

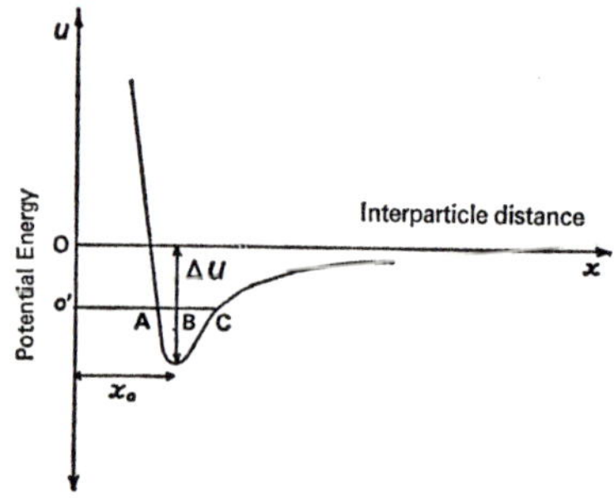

Figure 1.2b

crystalline solids and shows an important feature of such solids, namely, long-range order in the particle arrangement. The array obtained when all nearest neighbours have a common separation is shown for the two-dimensional case in fig. 1.3. In actual solids the equilibrium spacing of the particles is typically a few angstrom units (1Å $= 10^{-10}$m).

Fig. 1.3 is constructed on the assumption that only nearest-neighbour interactions are important, and that the attractive field of force round

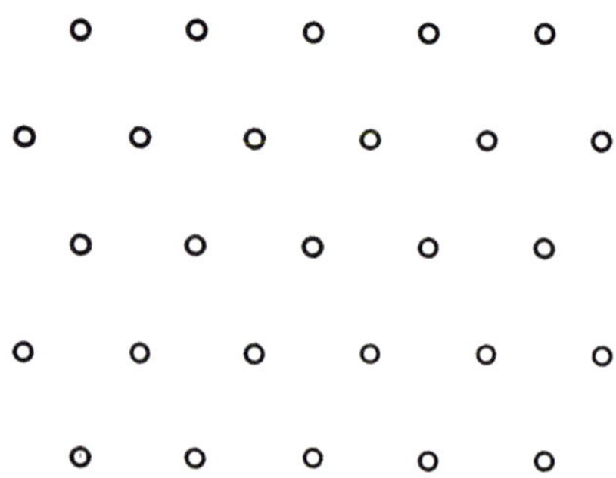

Figure 1.3

a particle is spherically symmetric. This would result in all crystals having the same particle arrangement, but with different values of x_0. However, the exact nature of the interparticle forces depends on the electron arrangements in the particles, and in most cases interparticle forces are not spherically symmetric. These departures from spherical symmetry influence the equilibrium positions taken up by the particles, so that three-dimensional arrays of different symmetries are obtained.

Thus, in general, though crystals of pure substances will be homogeneous, they will not be isotropic with respect to those properties that depend on the interparticle forces; for example, their elastic properties.

The predominant importance of nearest-neighbour interactions is similar to the concept of the chemical bond, in which a link is formed between adjacent particles, and it is often convenient to describe the particle interactions in terms of chemical bonds.

The existence of long-range order means that, in the undistorted crystal, axes can be set up that bear a definite relationship to the characteristic symmetry features of the particle arrangement. These are the crystallographic axes. In general, a solid can be divided into regions in each of which the crystallographic axes have an orientation different from those in adjacent regions. Each region is termed a 'grain', and the contact zone between two differently oriented grains is a 'grain boundary'. The grain boundary is a zone, a few particle spacings thick, in which the regular arrangement is distorted, and the particles

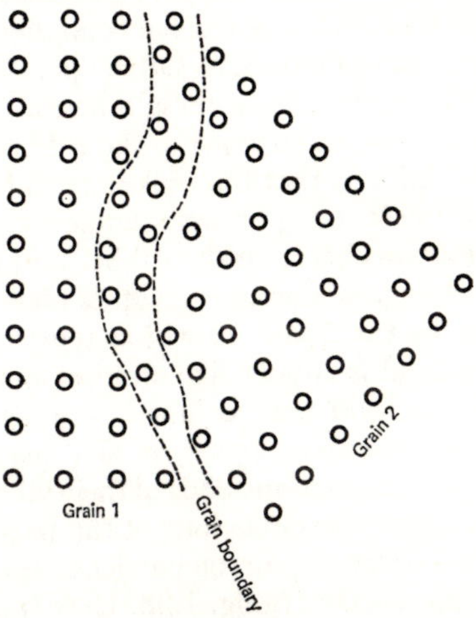

Figure 1.4

occupy equilibrium positions that are a compromise between the normal positions in the two lattices. A grain boundary is shown in two dimensions in fig. 1.4. If a specimen consists of a single grain it is called a 'single crystal'. Samples of crystalline solids often consist of a very large number of very small grains. In each grain the particles are arranged in a regular array, and the physical properties of the grain vary with direction. However, in general, the individual grains are arranged with random orientations of the crystallographic axes, so that the anisotropic effects are averaged out and the solid is effectively isotropic.

Most solids are crystalline, but not all. Some are amorphous and others partially crystalline. In amorphous solids there is no long-range order. The nearest-neighbour bonds are normal, but there is sufficient freedom in the interaction of adjacent groups of particles so formed that the long-range regularity is absent. A typical example is glass. The partially crystalline solids are those that have some regions in which long-range order is found and others in which it is absent. Many polymers in bulk show this property. Polymer molecules consist basically of very long chains of atoms and often contain small side groups of atoms at regular intervals along the chain. Such substances behave rather differently from 'ordinary' substances of low molecular weight. The earlier arguments of this chapter tacitly assumed that the particles could be treated as compact groups of atoms. To use the same ideas in the case of polymers, interactions between atoms, or groups of atoms, in any one molecular chain must be considered, and also those between adjacent chains.

Long-chain polymers can be crystalline in nature, though the crystallinity is usually rather imperfect. Some polymers are 'rubber-like'. In these the long molecule chains are very flexible and are arranged in an irregular tangled way. When a tension is applied to a sample of the material the chains straighten out. Other polymers are known as 'glassy' polymers. The molecules in these materials are in an irregular tangled configuration, but, in contrast to the rubber-like polymers, the molecule chains are rigid, so that the bulk material is hard and rigid.

Polymers in which cross-links between adjacent molecule chains are rare make transformations between the rubber and glassy conditions as the temperature is changed. The change takes place fairly sharply at a temperature known as the 'glass transition temperature'. Above this temperature the material is rubber-like and below it glassy.

If the particles in a solid possess kinetic energy they will oscillate about fixed positions, the kinetic energy in fact being totally converted into potential energy at the extreme ends of their vibrational movement. Assuming once more that the behaviour of the particles in the solid is similar to that of an isolated pair of particles, the situation may be represented by the line O'ABC in fig. 1.2b. Here it is assumed that one particle is fixed at O', and the other vibrates between the extreme positions A and C. The mean position of the particle is B, midway between A and C, so that the mean separation of the pair of particles is O'B. If the U–x curve is symmetrical about the minimum the particle separation will remain x_0, but if it is asymmetrical, the oscillation will be centred about some other position giving a greater separation, that is, thermal expansion, since the kinetic energy depends on temperature. As a result of fluctuations of kinetic energy, a few particles will be able to obtain sufficient kinetic energy to escape completely from the solid and give rise to the very low vapour pressure of a solid.

As the kinetic energy of the particles increases so does the amplitude of their oscillations. Even when this kinetic energy is less than ΔU, (see fig. 1.2b) they can at some stage break away from their neighbours, but are quickly recaptured by another group, so that short-range order is maintained, but long-range order is destroyed. This is the liquid phase, where the main motion of any particle is moving from one short-lived 'cell' to another. The volume change on melting is small, so that the change in U is small. The potential energy-separation curve for a pair of particles gives no information about the value of the melting-point.

If the kinetic energy increases, so that the mean value is greater than ΔU, the particles can all escape completely from their nearest neighbours. This process gives rise to the gaseous phase, in which the separations of the particles are sufficiently great for the interparticle forces to be ineffective, except during collisions.

Further reading

Christy, R. W., and Pytte, A., *The Structure of Matter*, W. A. Benjamin, Inc., 1965, ch. 14.

2
the elastic properties of matter

2.1 Elasticity

The discussion in this chapter applies to the behaviour of crystalline solids, though many of the results are also true for non-crystalline 'ordinary' materials. The behaviour of polymers is excluded, since they frequently show deviations from common laws such as Hooke's Law (see section 2.3). Also, the load needed to produce a given deformation in a polymer is not usually a single-valued function of the deformation, but depends on the previous history of the specimen. When a polymer specimen is loaded, there is an almost instantaneous extension, followed by a slow creep which may persist for a very long time. On unloading, there is an almost immediate recovery of some of the extension, some is recovered in time and some is never recovered.

When a crystalline solid is acted upon by a system of applied forces, these forces try to displace the particles from their equilibrium positions. This tendency is opposed by restoring forces which are set up as soon as the displacement occurs. If tensile deforming forces are applied that try to increase the particle separation in the solid, the restoring forces are attractive; if the deforming forces are compressive, the restoring forces are repulsive. The restoring force builds up with the displacement of the particles from their equilibrium positions to a value that balances the applied force, giving an equilibrium state, and is maintained as long as the applied force is maintained. When the deforming forces are removed, the restoring forces will bring the particles back to the equilibrium positions they occupied before the application of the external forces. This property of solids (and, in some circumstances, of fluids) to suffer a deformation under an applied load, and then recover their original shape completely on removal of the load, is termed elasticity.

In the immediate neighbourhood of the equilibrium position E on the interparticle force-interparticle distance graph (repeated as fig. 2.1) the curve may be treated as a straight line. Then, in this region, if the separation of the particles is changed slightly (either under tension or compression) the restoring force is proportional to the displacement, that is, considering the bulk crystal, the applied force is proportional to the deformation it produces. As the applied force is further increased at constant temperature, the deformation may still

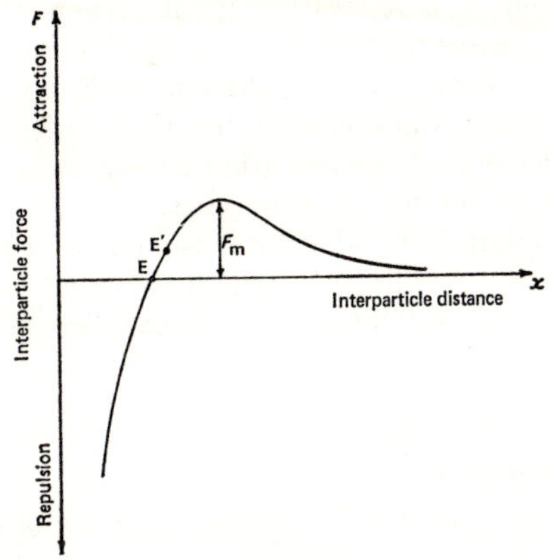

Figure 2.1

be elastic though the relation between applied force and deformation is no longer linear.

Fig. 2.1 applies strictly to a particle with one neighbour only. For small changes in the separation of two such particles, the restoring force F may be written

$$F = -\alpha \Delta x$$

where Δx is the change in particle separation, and α is a constant known as the 'force constant' for the particles concerned.

Consider a single row of identical particles—the one-dimensional solid—in which only nearest neighbour inetractions are important. Let one particle be displaced by a small distance Δx, the others remaining fixed. Since only nearest neighbour interactions are important, the forces exerted on the displaced particle by its neighbours can be considered separately. The restoring force acting on the displaced particle due to the fixed particle to the right is $-\alpha \Delta x$, and that due to the fixed particle to the left is also $-\alpha \Delta x$. Thus, the net restoring force is $-2\alpha \Delta x$, and the equation of motion of the displaced particle is

$$\text{acceleration} = -\frac{2\alpha}{m} \Delta x$$

where m is the mass of the particle. This equation represents a simple harmonic motion, with a frequency $\nu = \frac{1}{2\pi}\sqrt{\frac{2\alpha}{m}}$, which is thus the natural frequency of vibration of the solid and is often referred to as the Einstein frequency.

The particle separation is x_o only in the absence of particle kinetic energy. Because of the asymmetry of the U–x curve, when the kinetic energy is other than very small, the positions about which the particles vibrate correspond to a separation greater than x_o, as explained in section 1.2. This means that when using fig. 2.1 to deduce mechanical properties, the working point should not be E but E$'$. At this point the slope of the curve is numerically less than at E, and thus the force needed to produce a given deformation decreases as the kinetic energy increases, that is, as the temperature rises.

Fig. 2.1 shows that, if no other processes occur first, when the restoring force reaches F_m the force to produce further increase in particle separation decreases, and catastrophic separation of the particles will occur.

2.2 Stress and strain

When a body has external forces applied to it, or when one part of the body exerts a force on a neighbouring part, the body is said to be in a state of stress. Two types of forces may be acting on any element of a body (*a*) 'body forces' which are proportional to the volume of the element, e.g. gravitational forces, and (*b*) 'surface forces' exerted by the surrounding material and which are proportional to the surface area of the element. In most cases the effect of body forces is negligible compared with that of surface forces.

The 'stress' in an element is measured by the quantity $\frac{\text{force}}{\text{area}}$ and thus has SI units N m^{-2}. The stress in a body is described as 'homogeneous' if, for an element of fixed shape and orientation, the value is independent of the position of the element in the body.

The forces applied to the surface of an element, and giving rise to the state of stress, may make any angle with the surface of the element. Consider a force F acting uniformly over a plane surface of area A, and making an angle θ with the normal to the area, as in fig. 2.2a. The component of the force, per unit area, acting in the plane is termed the 'tangential' or 'shearing' stress τ, and

$$\tau = \frac{F \sin\theta}{A}$$

The component acting normal to the plane is the 'contractile' or 'tensile' stress, depending respectively on whether its effect is to decrease or increase the particle separation, and is denoted by σ.

$$\sigma = \frac{F\cos\theta}{A}$$

In the case where the surface is not plane, or where the force is not uniformly distributed over the surface, the term stress is usually taken

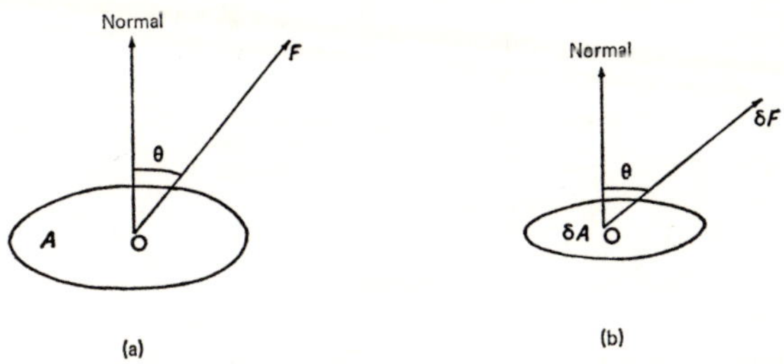

Figure 2.2

to mean 'stress at a point', which is the limiting value of $\frac{\text{force}}{\text{area}}$ as the area tends to zero. If a force δF acts over an area δA, and makes an angle θ with the normal to the area as in fig. 2.2b, the normal stress at O is

$$\sigma = \underset{\delta A \to 0}{\mathrm{Lt}} \left(\frac{\delta F \cos\theta}{\delta A} \right)$$

The application of a single force to a body produces a linear acceleration in that body, while the application of a couple produces an angular acceleration. For a body to be in a state of stress and in statical equilibrium, all forces and couples must be balanced by equal and opposite ones. Thus, a bar in tension must be acted upon by equal and opposite forces acting along the same axis, as in fig. 2.3, and a body acted upon

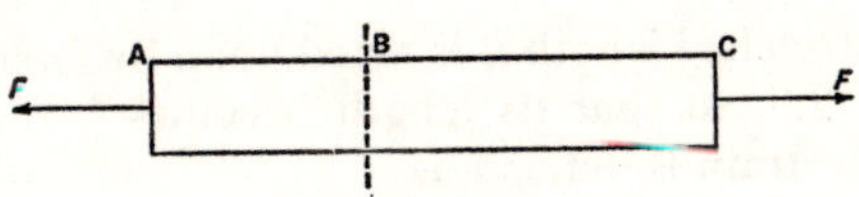

Figure 2.3

solely by shear stresses must be in equilibrium under the effect of two equal and opposite couples, as in fig. 2.4.

These examples show that when forces are applied to the surfaces of a body the stresses act through the whole of the body in equilibrium. Consider the equilibrium of any length, e.g. BC, of the bar of fig. 2.3. The forces exerted by AB on BC must maintain the equilibrium with the force acting over C. The applied force is superimposed on the inter-particle force distribution, and changes the particle separation until the restoring force set up exactly balances the applied force.

This change in particle separation is seen on the macroscopic scale

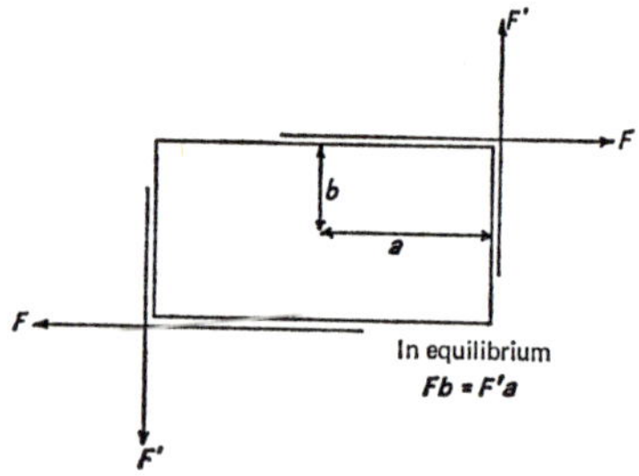

Figure 2.4

as a change in the dimensions of the body that is stressed. The change in size is characterized by the strain, which may be defined as

$$\text{strain} = \frac{\text{change in dimension}}{\text{original dimension}}$$

Strain is thus a ratio and has no units. As defined above, strain measures the change in the relative positions of the basic particles of a substance. It should be noted that, if a body is displaced, or undergoes rotation as a rigid body, it suffers no strain. The strain is described as 'homogeneous' if, after deformation, lines on the body that were originally straight remain straight, and lines that were originally parallel remain parallel.

Three basic strains are found to be convenient in describing the behaviour of a body in various states of stress. These are as follows:

(*a*) *Tensile strain*

If a body of unstretched length l_o is acted upon by forces applied at the ends, as in fig. 2.3, so that its length becomes l, the 'conventional' (tensile or axial) strain is defined as

$$\epsilon = \frac{l - l_o}{l_o}$$

Sometimes it is more convenient to use a 'logarithmic' or 'natural' (tensile) strain, which is defined as

$$\epsilon^* = \ln(l/l_o)$$

$$\text{Clearly } \epsilon^* = \ln(1+\epsilon)$$

When the extension is small $\epsilon \simeq \epsilon^*$

(*b*) *Bulk strain*

If, as a result of an applied uniform hydrostatic force, a solid of initial volume V_o changes to one of volume V the bulk strain θ is defined as

$$\theta = \frac{V - V_o}{V_o}$$

(*c*) *Shear strain*

Let a cube of material be sheared through an angle γ by applied couples as shown in fig. 2.5. The angle γ is here taken as the measure of the

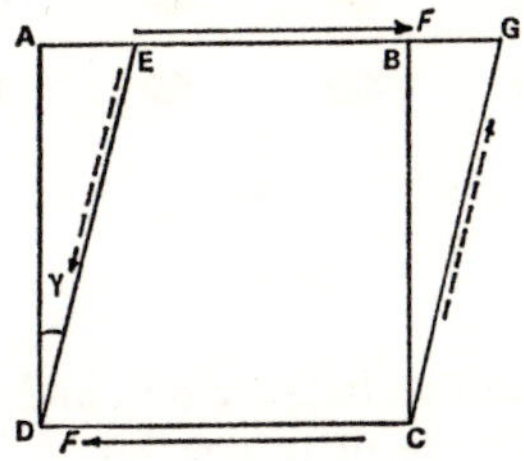

Figure 2.5

strain, in this case a shear strain. When γ is small the shear strain is approximately AE/AD.

2.3 The relationship between stress and strain

When the strain produced in a solid, as the result of a stress acting upon it, is small, it is found experimentally that

$$\text{strain} \propto \text{stress},$$

or,

$$\text{stress} = \text{constant} \times \text{strain}.$$

This is Hooke's Law and was first stated in 1660 in connection with the extension of a loaded spring. The constant in the equation as it is written is called a 'modulus of elasticity'. The graph of stress against strain for such small deformations is a straight line passing through the origin. It has already been shown that this linear relationship is expected for small distortions, that is, for small changes in the interparticle distance, because a short length of the interparticle force–interparticle distance curve, in the neighbourhood of the equilibrium distance, can be treated as a straight line, whether or not the effect of particle kinetic energy is taken into account.

In a crystal the law of force between the particles will, in general, depend on the orientation. For example, if fig. 2.6 represents a section of a crystal, and if, for the particles of this crystal, the slope at E (fig. 2.1) depends on orientation, the tensile stress to produce a given strain in the direction AA will not be the same as that needed to produce the same strain in the direction BB. Thus, the modulus of elasticity for a single crystal will, in general, depend on the direction in which it is measured. This is true for the individual grains in a polycrystalline sample, but if the orientations of the grains are random, and if the grain size is small

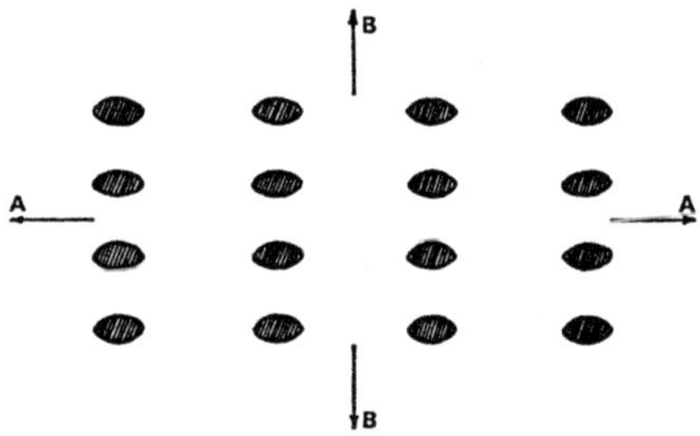

Figure 2.6

enough, the orientation effects balance out, and the sample is effectively isotropic with respect to its elastic properties.

It has been shown in section 2.1 that the force needed to produce a given change in particle separation is temperature dependent, and thus the modulus of elasticity will be temperature dependent and will in most cases decrease as the temperature increases.

2.4 The moduli of elasticity

The three strains commonly used to describe the deformation of stressed bodies give rise to three corresponding moduli of elasticity. If elastically isotropic solids are considered, the moduli are independent of the direction of their measurement in the solid.

Consider an isotropic bar of uniform cross-section A and unstretched length l_0, acted upon by forces F applied uniformly at the ends, as in fig. 2.3. If l is the length under the applied load, the tensile strain will be homogeneous and equal to $\frac{l-l_0}{l_0}$

The stress σ, which is homogeneous also, is given by

$$\sigma = \frac{F}{A}$$

if the convention is adopted that tensile stresses are counted positive, and compressive stresses negative. Thus, when Hooke's Law is obeyed

$$\sigma = \frac{F}{A} = E\left(\frac{l-l_0}{l_0}\right) \tag{2.1}$$

where E is a constant at a particular temperature, and in this case is known as Young's modulus.

In the neighbourhood of the equilibrium distance, the interparticle force–interparticle distance curve is effectively a straight line, and hence the value of the Young's modulus for a given material is the same in tension and compression.

The value of Young's modulus for a given material can be used to obtain information regarding the force constant (see section 2.3) for that material. For example, consider a cubic structure in which the spacing between planes of particles, and also between particles in a row, is a_0. Unit area measured perpendicular to the rows of particles contains $1/a_0^2$ rows, so that each row, of n particles, say, may be said to have a cross-sectional area of a_0^2. Let a force F be applied in the direction of such a row, producing a change in a_0 of Δa_0. Then, applying Hooke's Law to this row of particles,

$$\frac{F}{a_0^2} = \frac{En\Delta a_0}{na_0} = \frac{E\Delta a_0}{a_0}$$

or,

$$F = Ea_0\,\Delta a_0$$

The force exerted by the particles is $-F$, and as this is transmitted along the row of particles, this will be the force acting on a particle when the particle separation is changed by Δa_0. It has been shown that this is equal to $-2\alpha\,\Delta a_0$, and hence the force constant is $\alpha = Ea_0/2$. Typically E is of the order of 10^{11} N m^{-2} and a_0 is of the order of $0{\cdot}5 \times 10^{-9}$m, giving a typical value for α of 25 N m.

When a cube of material is subjected to a uniform hydrostatic stress, as a result of forces F being applied in the same sense to each surface of area A, a change in volume from V_0 to V is produced. The stress σ is F/A and the bulk strain is $\dfrac{V-V_0}{V_0}$. Thus, when Hooke's Law is obeyed,

$$\sigma = \frac{F}{A} = K\left(\frac{V-V_0}{V_0}\right) \tag{2.2}$$

where K is a constant at constant temperature, called the bulk modulus. This modulus is used when a solid is subjected to an applied hydrostatic pressure. A pressure has a negative sign, but gives rise to a decrease in volume, so that K in equation (2.2) has a positive sign.

When a cube of material is sheared as a result of applied couples, produced by forces F acting tangentially over a surface of area A, as shown in fig. 2.5, the angle of shear γ is taken as a measure of the strain. When Hooke's Law is obeyed

$$\tau = \frac{F}{A} = G\gamma \tag{2.3}$$

where G is a constant at constant temperature known as the rigidity modulus.

In each case, the modulus has a single value at a given temperature for a particular material, only if the material can be treated as elastically isotropic.

2.5 Poisson's ratio

When a prismatic bar with free (that is, unloaded) prismatic surfaces is deformed by an axial load, in addition to a change in length, a change in the cross-sectional area is produced. Experiment shows that the ratio

$$\nu = \frac{\text{change in lateral dimension/original lateral dimension}}{\text{change in length/original length}}$$

is a constant when the deformation is elastic, and is called Poisson's ratio.

This change in cross-section with extension necessitates a more careful definition of stress in this case. Two definitions are in common use:

(*a*) *Conventional stress*

This is defined as

$$\frac{\text{load at any extension}}{\text{original area of cross-section}}$$

Since the original area of cross-section is a constant, this stress is directly proportional to the applied load.

(b) True stress

This is defined as

$$\frac{\text{load at any extension}}{\text{area of cross-section under that load}}.$$

These two stresses are approximately equal for small strains.

2.6 Some typical values

Some typical values of the elastic moduli for ordinary laboratory materials are given below. They refer to room temperature measurements on isotropic (fine-grained) materials.

Table 2.1

Material	E N m^{-2}	K N m^{-2}	G N m^{-2}	ν
Aluminium	71×10^9	75×10^9	26×10^9	0·33
Copper	130×10^9	138×10^9	46×10^9	0·34
Steel	210×10^9	168×10^9	83×10^9	0·28

The temperature dependence of Young's modulus can be expressed as a temperature coefficient α defined by the equation

$$E_t = E_{15}(1-\alpha(t-15))$$

where E_t is the value of Young's modulus at t °C and E_{15} is its value at 15°C. Values of α are given below.

Table 2.2

Material	α (deg^{-1})
Aluminium	$4{\cdot}8\times10^{-4}$
Copper	$3{\cdot}0\times10^{-4}$
Steel	$2{\cdot}4\times10^{-4}$
Quartz fibre	$-1{\cdot}5\times10^{-4}$

2.7 The theorem of superposition

This theorem states that when strains are small (as is usually the case when the behaviour of a solid is perfectly elastic) the effects of different stresses are additive.

As an example of the application of the theorem, consider a cube of isotropic material with edges of length unity, each face of which is subject to a tensile force σ. The normal stress over each face is then σ. Using the theorem of superposition, the procedure is to find the effect of each stress acting separately, and then obtain the total effect of all the stresses acting simultaneously by adding together the separate effects.

Set up orthogonal axes Ox, Oy, Oz parallel to the cube axes and let the applied stresses be σ_x, σ_y, σ_z, respectively.

(i) The effect of stress σ_x acting alone.

σ_x produces a strain ϵ_{x1} in the direction Ox, given by

$$\epsilon_{x1} = \frac{\sigma_x}{E}$$

It also produces contractions along Oy and Oz, giving strains ϵ_{y1} and ϵ_{z1}, respectively, where

$$\epsilon_{y1} = -\nu\epsilon_{x1} \text{ and } \epsilon_{z1} = -\nu\epsilon_{x1}$$

ν is Poisson's ratio and E is Young's modulus.

(ii) Similarly, the effect of stress σ_y acting alone is to produce an extension in the direction Oy for which the strain is

$$\epsilon_{y2} = \frac{\sigma_y}{E}$$

accompanied by transverse strains

$$\epsilon_{x2} = -\nu\epsilon_{y2} \text{ and } \epsilon_{z2} = -\nu\epsilon_{y2}$$

(iii) Again, the effect of stress σ_z acting alone is to produce the following strains

$$\epsilon_{z3} = \frac{\sigma_z}{E};\ \epsilon_{y3} = -\nu\epsilon_{z3};\ \epsilon_{x3} = -\nu\epsilon_{z3}$$

Using the theorem of superposition, the total strain along Ox is

$$\epsilon_x = \epsilon_{x1} + \epsilon_{x2} + \epsilon_{x3}$$

Then,
$$\epsilon_x = \frac{\sigma_x}{E} - \nu\epsilon_{y2} - \nu\epsilon_{z3}$$

$$= \frac{\sigma_x}{E} - \frac{\nu\sigma_y}{E} - \frac{\nu\sigma_z}{E}$$

$$= \frac{\sigma}{E}(1-2\nu)$$

since in this case

$$\sigma_x = \sigma_y = \sigma_z = \sigma$$

Similarly, the total strain along Oy is

$$\epsilon_y = \frac{\sigma}{E}(1-2\nu)$$

and that along Oz is

$$\epsilon_z = \frac{\sigma}{E}(1-2\nu)$$

If second order terms such as $\epsilon_x\epsilon_y$ are neglected, the dilatation, that is the change in volume, is, since the cube edges are of unit length,

$$\epsilon_x + \epsilon_y + \epsilon_z = \frac{3\sigma}{E}(1-2\nu)$$

and is in this case the change in volume per unit volume, that is, the bulk strain.

This is equal to σ/K, where K is the bulk modulus, and hence

$$\frac{3}{E}(1-2\nu) = \frac{1}{K}$$

or,
$$E = 3K(1-2\nu) \tag{2.4}$$

This equation shows that ν cannot be greater than $\frac{1}{2}$, since E and K must be positive.

Now take a unit cube of isotropic material subjected to the forces shown in fig. 2.7. Considering the effect of each stress acting separately, and then using the theorem of superposition to obtain the effect of all the forces acting simultaneously, it follows that

$$\epsilon_x = \frac{\sigma}{E} + \frac{\nu\sigma}{E} = \frac{\sigma}{E}(1+\nu)$$

$$\epsilon_y = -\frac{\sigma}{E}(1+\nu)$$

$$\epsilon_z = 0$$

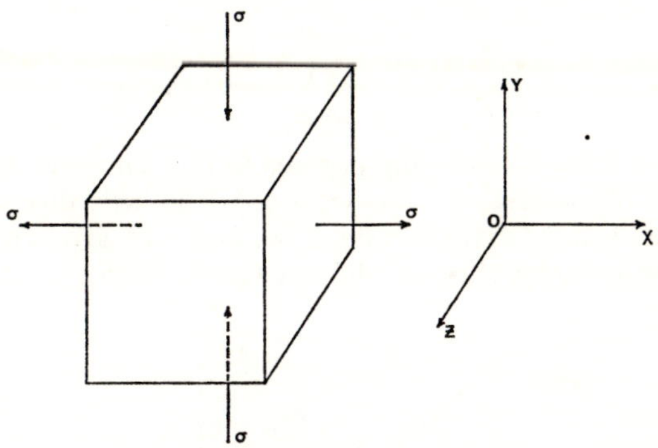

Figure 2.7

Consider now a section through the cube, parallel to the x–y plane, shown in fig. 2.8a. Since EFGH is a section of a unit cube

$$BF = FC = 1/2$$

and

$$BC = 1/\sqrt{2}$$

Next, consider the equilibrium of the prism of material with cross section BFC. This is shown separately in fig. 2.8b. The applied stresses acting on the surfaces represented by BF, FC must be in equilibrium

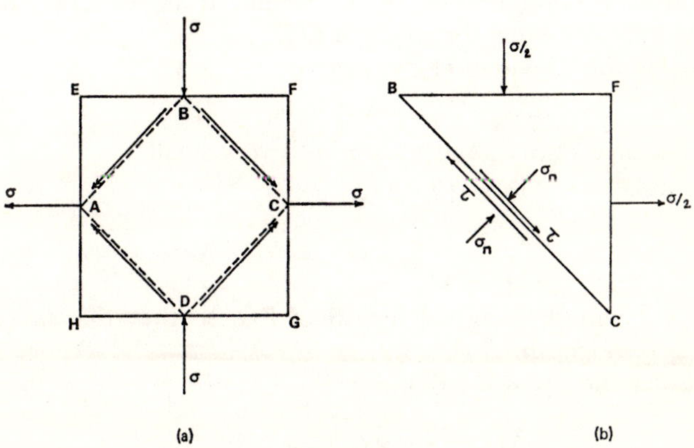

Figure 2.8

with those acting across BC which arise from the presence of the remainder of the stressed solid.

Resolving forces normal to BC gives

$$\frac{\sigma_n}{\sqrt{2}} = \frac{\sigma}{2\sqrt{2}} - \frac{\sigma}{2\sqrt{2}} = 0$$

where σ_n is the stress normal to BC.

Resolving forces parallel to BC gives

$$\frac{\tau}{\sqrt{2}} = \frac{\sigma}{2\sqrt{2}} + \frac{\sigma}{2\sqrt{2}} = \frac{\sigma}{\sqrt{2}}$$

Hence $$\tau = \sigma$$

where τ is the shear stress acting parallel to BC. Thus, the forces acting on the prism of material of cross-section ABCD have directions shown by the double-headed arrows in fig. 2.8a, and each gives rise to a stress of magnitude σ. Such a state of affairs is a shear stress.

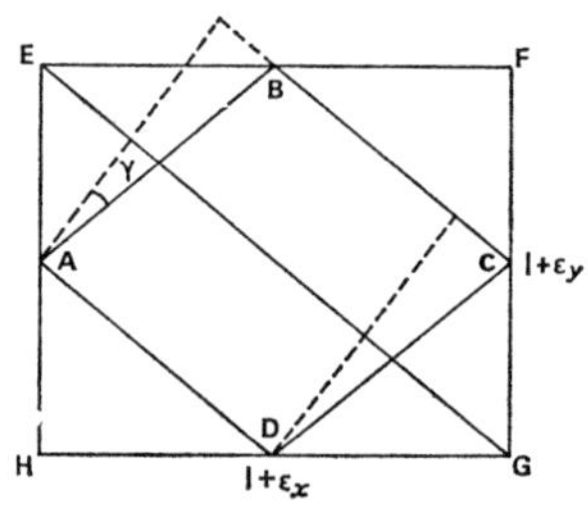

Figure 2.9

In the example considered the distorted shape of the cube is really as in fig. 2.9. This distortion was neglected in the calculation of the stress. From a consideration of the similar triangles EFG, BFC the strain of BC must equal the strain of EG.

Now, by the Theorem of Pythagoras,

$$EG = [(1+\epsilon_x)^2+(1+\epsilon_y)^2]^{\frac{1}{2}}$$

If it is assumed that ϵ_x and ϵ_y are both very small

$$EG \simeq [2+2\epsilon_x+2\epsilon_y]^{\frac{1}{2}}$$

and, since $$\epsilon_x = -\epsilon_y$$

$$EG \simeq \sqrt{2}$$

But $\sqrt{2}$ was the original length of EG, so that EG and BC are unchanged in length.

Consider the triangle ABD.

$$B\hat{A}D = \frac{\pi}{2} - \gamma$$

since its original value was $\pi/2$ and the square ABCD has suffered a shear strain γ. Also,

$$(\text{EH})^2 = (\text{BD})^2 = (1+\epsilon_y)^2 = (\text{AB})^2+(\text{AD})^2-2\text{AB.AD}\cos\text{B}\hat{\text{A}}\text{D}$$
$$= \frac{1}{2}+\frac{1}{2}-2\frac{1}{\sqrt{2}}\times\frac{1}{\sqrt{2}}\sin\gamma$$

If it is assumed that both ϵ_y and γ are very small, this gives

$$1+2\epsilon_y = 1-\gamma$$

so that

$$\gamma = -2\epsilon_y = 2\epsilon_x$$

Now $\dfrac{\tau}{\gamma} = G$, the rigidity modulus, so that

$$G = \frac{\tau}{\gamma} = \frac{\sigma}{2\epsilon_x} = \frac{\sigma E}{2\sigma(1+\nu)} = \frac{E}{2(1+\nu)} \tag{2.5}$$

Thus, for an isotropic solid

$$E = 3K(1-2\nu)$$

and

$$G = \frac{E}{2(1+\nu)}$$

This section has shown, not only how the theorem of superposition is used, but also that the four elastic constants for a particular material, E, K, G and ν are not independent of each other. Any one of them can be expressed in terms of any two of the others.

2.8 Strain energy

Work must be done by an external force on an elastic solid in order to deform the latter, and this work is stored as potential energy in the distorted solid. If the deformation is purely elastic, the whole of this energy can be recovered when the force producing the deformation is removed.

The increment of work δW done when a constant applied force F moves its point of application a distance δe is

$$\delta W = F\cos\theta\,\delta e$$

where θ is the angle between the direction of the force and that of the displacement it produces.

If work is performed by a constant couple of moment T causing a rotation through an angle $\delta\varphi$, the increment of work done by the couple is

$$\delta W = T\delta\varphi$$

Consider a rod deformed in tension. If F is the applied force and this produces an axial extension e, the work that must be done to produce a further extension de, an infinitestimal change, is the limiting case of the earlier expression, namely

$$dW = F\,de.$$

In this case θ is zero. Let F be a function of e that can be represented graphically as in fig. 2.10. Then, as the applied force is increased from F_1 to F_2 the extension changes from e_1 to e_2 and the work that must be done on the specimen is

$$W = \int_{e_1}^{e_2} F\,de = \text{shaded area under graph.}$$

No conditions have been placed on the dependence of e on F, and all the stored energy is recoverable, provided the behaviour is perfectly elastic.

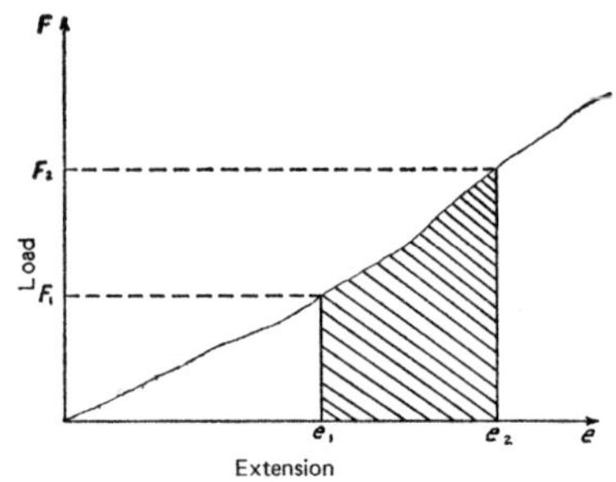

Figure 2.10

If Hooke's Law is obeyed and the rod has an unstretched length l_o and area of cross-section A, then, since

$$\sigma = E\epsilon$$

$$F = \sigma A$$

and

$$e = \epsilon l_o$$

where σ is the stress, ϵ the strain and e the extension, the expression for W becomes

$$W = \int_{e_1}^{e_2} \frac{AEe}{l_o}\,de$$

$$= \frac{AE(e_2^2 - e_1^2)}{2l_o}$$

$$= \frac{AEl_o(\epsilon_2^2 - \epsilon_1^2)}{2} \qquad (2.6)$$

If volume changes are assumed to be small, the volume is Al_o and hence the energy stored per unit volume is

$$\frac{E(\epsilon_2^2 - \epsilon_1^2)}{2}$$

When ϵ_1 is zero the expression for the energy stored per unit volume can be written

$$\frac{E\epsilon_2^2}{2} = \frac{\sigma_2^2}{2E} = \frac{\sigma_2\epsilon_2}{2}$$

2.9 The torsion of a cylinder

Consider a solid, isotropic, right circular cylinder of length l and radius a, with one end twisted relative to the other through an angle φ radians, as in fig. 2.11a. ABCDEFGH is the undistorted shape of an element of the cylinder at a distance r from the axis (fig. 2.11b). Let ABCDJKLM be the shape of the element after deformation. The base of the cylinder is shown in fig. 2.11c.

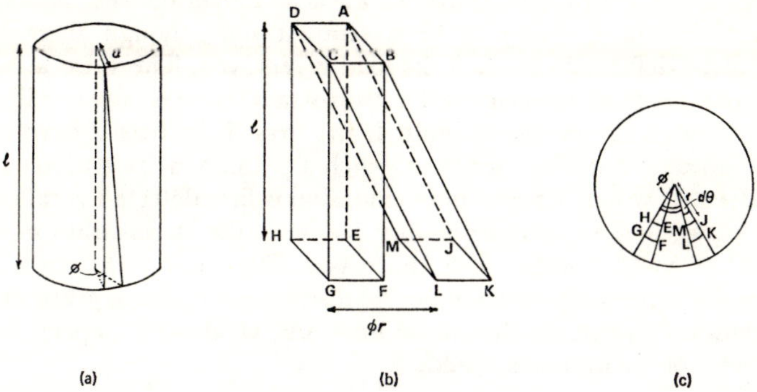

Figure 2.11

The area of the element is $r\,d\theta\,dr$ and the shear strain it suffers is $\varphi r/l$, where $d\theta$ is the angle subtended at the axis by the element. The applied force needed to produce this strain is

$$\text{stress} \times \text{area} = \text{strain} \times \text{modulus} \times \text{area}$$

$$= \frac{\varphi r}{l} \times G \times r\,d\theta\,dr$$

where G is the rigidity modulus. This force has a moment about the axis given by

$$\text{force} \times \text{distance from axis}$$

$$= \frac{\varphi r^2 G\,d\theta\,dr}{l}\,r$$

Hence, the total moment, which is equal to the couple that must be applied at each end of the cylinder to give equilibrium, is

$$T = \int_o^{2\pi} d\theta \int_o^a \frac{\varphi r^3 G}{l}\,dr$$

$$= \frac{\pi G a^4}{2l}\phi \qquad (2.7)$$

If the cylinder is hollow, having an internal radius a_o and an outer radius a,

$$T = \frac{\pi G(a_1{}^4 - a_o{}^4)}{2l}\ \varphi \tag{2.8}$$

The quantity T/φ is called the torsional rigidity.

2.10 Elastic limit and yield point

When a solid is deformed in tension under an increasing stress, a stage is reached when the further deformation produced by a slight increase in stress is still elastic, but Hooke's Law is no longer obeyed, that is, the stress-strain curve ceases to be a straight line, though there is no permanent deformation when the load is removed. The stress at which the departure from linearity of the stress-strain curve occurs is called the elastic limit or proportional limit. If the stress is increased beyond the elastic limit, a value is reached at which a permanent extension occurs. The stress at which a permanent extension is first detected is called the yield stress, and the corresponding point on the stress-strain curve is called the yield point. Often the elastic limit and yield stress have almost the same value and cannot be distinguished. For a given sample the values obtained for the elastic limit and yield stress depend on the sensitivity of the apparatus used.

2.11 The elasticity of fluids

Since a fluid cannot sustain a shear stress, the only modulus that can be defined is the bulk modulus,

$$K = \frac{\sigma V_o}{V - V_o} \tag{2.2}$$

The compressibility β is defined as $1/K$.

When changes in hydrostatic pressure are made slowly, the fluid remains at constant temperature if it is in a conducting vessel, and the corresponding modulus is termed the 'isothermal' bulk modulus. On the other hand, if changes occur rapidly, so that there is no time for any significant heat exchange with the surroundings, the 'adiabatic' bulk modulus is obtained.

In the case of measurements on liquids, the major problems are that β is small, so that large pressure changes are needed to produce quite small volume changes, and, since the containing vessel is also affected, its change in volume must be taken into account.

2.12 The tensile strength of liquids

Under certain conditions liquids are able to sustain large tensions. One way in which this can be demonstrated is due to Berthelot. A cylindrical tube is almost completely filled with liquid, the remainder of the space being air and liquid vapour, and is then sealed. When the tube is heated, the liquid expands faster than the tube and forces the air into solution, so that, at a certain temperature T_f, the liquid fills the tube completely. The tube is then allowed to cool. During this cooling the liquid adheres to the walls of the tube, and continues to fill it completely until a temperature T_b, which is less than T_f, is reached, when the liquid ruptures as a result of the tension set up in it due to cooling. At the instant at which rupture occurs the release of tension is accompanied by a 'click', and a sudden increase in the volume of the Berthelot tube.

The nature of the walls of the tube has an important effect on the results obtained in the above experiment, and it is likely that what is being measured is the adhesion of the liquid to the solid, rather than the true liquid tensile strength. Using a glass tube, Temperley and Chambers showed that water could withstand tensions of between 3 and 5×10^6 N m^{-2}. Briggs has used a different method to show that the tension to cause rupture is strongly temperature-dependent for water in a glass tube, and obtained a maximum tension of about 28×10^6 N m^{-2} at 10°*C*.

2.13 Examples

(*a*) One half of a rod of length $2l$ has a radius r and the other half has a radius $2r$. The smaller end is clamped while the other is twisted about the axis. Find the torsional rigidity of the complete rod if the rigidity modulus of the material, assumed isotropic, is G.

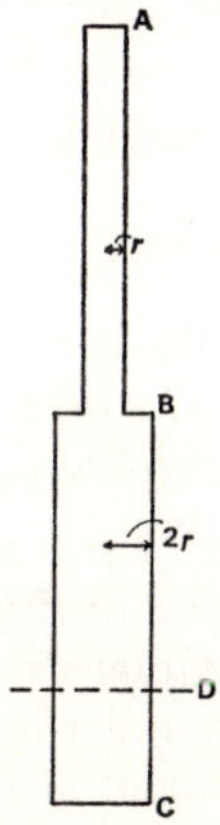

Figure 2.12

Calculate the couple needed to twist the larger end of the rod through 5° relative to the other if the rod is 0·2m long, the larger radius is $2{\cdot}0 \times 10^{-3}$m and the rigidity modulus is 40×10^9N m^{-2}.

Let the twisted rod be cut into two on any plane normal to its axis, e.g. on D (see fig. 2.12). To maintain the twisted portions in equilibrium, a couple must be applied to the end D of DC equal to that exerted on DC by the portion of the bar AD before the cut was made. This couple must equal the applied couple acting at C. Clearly this result holds for any position of the cut.

Let the applied couple be T, the twist produced at B relative to A be φ_B and that at C relative to A be φ_C.

Then, applying the basic equation to AB and to BC

$$T = \frac{\pi G(2r)^4(\varphi_C - \varphi_B)}{2l} \qquad \text{for BC.}$$

$$T = \frac{\pi G r^4 \varphi_B}{2l} \qquad \text{for AB.}$$

Thus,
$$\frac{\varphi_B}{\varphi_C - \varphi_B} = 16$$

or,
$$16\varphi_C = 17\varphi_B$$

The torsional rigidity of the complete rod is $\frac{T}{\varphi_C}$, and this is given by

$$\frac{T}{\varphi_C} = \frac{\pi G r^4}{2l} \times \frac{16}{17}$$

$$= \frac{8\pi G r^4}{17\,l}$$

For a twist of 5° using the values given

$$T = \frac{8\pi \times 40 \times 10^9 \times (10^{-3})^4 \times \varphi_C}{17 \times 0{\cdot}1}$$

For the couple to be in N m, φ_C must be in radians.

$$5° = \frac{\pi}{36} \text{ radians.}$$

Hence,
$$T = \frac{8\pi \times 40 \times 10^9 \times 10^{-12} \times \pi}{17 \times 0{\cdot}1 \times 36}$$

$$= 51{\cdot}5 \times 10^{-3} \text{N m}$$

(*b*) A wire of unstretched length l and area of cross-section A, made of isotropic material of density ρ and having a Young's modulus E is freely suspended at one end. Find the total strain in the wire and the potential energy stored due to stretching under its own weight.

Take the origin of measurement at the lower end of the suspended wire, and consider an element of length dy at a distance y from the origin. This element is acted on by a force due to the weight of the wire between the origin and y, that is, by a force $\rho g A y$, where g is the intensity of gravity. The stress in the element resulting from this force is $\rho g y$, and this gives rise to a strain in the element of

$$\frac{\rho g y\, dy}{E}$$

Hence, the total strain is

$$\int_0^l \frac{\rho g y\,dy}{E} = \frac{\rho g l^2}{2E}$$

The elastic energy stored in the element is $\frac{1}{2}$ load $\times$ extension, which equals $\frac{1}{2}\,\frac{(\text{stress})^2\ \text{volume}}{E}$, that is,

$$\frac{\frac{1}{2}(\rho g y)^2 A\,dy}{E}$$

Hence the total energy stored in the wire is

$$\int_0^l \frac{\rho^2 g^2 A y^2\,dy}{2E} = \frac{\rho^2 g^2 A l^3}{6E}$$

Problems

Assume linear elastic behaviour in all cases.

1. The axial modulus of elasticity for a solid is defined as

$$\chi = \frac{\text{longitudinal load per unit area of cross-section}}{\text{increase in length per unit length}}$$

subject to the condition that there is no change in area of cross-section during deformation, that is, lateral stresses must be applied to prevent lateral contraction. Show that, for an isotropic solid,

$$\chi = \frac{3K+4G}{3}$$

where K is the bulk modulus and G the rigidity modulus.

2. A metal tube 2 m long and of internal and external radii 0·040 m and 0·043 m respectively is clamped at one end. Calculate the magnitude of the couple which must be applied to the other end to twist it through 30°. The rigidity modulus of the metal, which is isotropic, is 25×10^9 N m^{-2}.

3. A uniform wire suspended vertically from a rigid support extends $0{\cdot}5 \times 10^{-3}$ m when a load of 1 kg is applied to the lower end. Calculate the elastic energy stored in the wire and the change in potential energy of the load. Comment on the result.

4. A uniform wire of circular cross-section has an unstretched length of 2 m and radius $1{\cdot}0 \times 10^{-3}$ m. The material is isotropic with Young's modulus 81×10^9 N m^{-2}. The wire is suspended vertically from a rigid support and a load of 1·5 kg is applied to the lower end. Calculate the change in elastic energy stored in the wire when the load is increased by a further 1 kg.

5. Show that in the neighbourhood of the equilibrium position of a particle in a solid, the potential energy—interparticle distance curve is parabolic.

6. The force constant for copper atoms is 23 N m^{-1} at room temperature. Estimate the natural frequency of the copper lattice under these conditions. Atomic weight of copper is 63·5; mass of the hydrogen atom is $1 \cdot 68 \times 10^{-27}$ kg.

Further reading

Christy, R. W., and Pytte, A., *The Structure of Matter*, W. A. Benjamin, Inc., 1965, ch. 12–14.

Temperley, H. N. V., *Properties of Matter* (third edition), University Tutorial Press, 1965, ch. 4.

3
the permanent deformation of solids

3.1 Introduction

In the previous chapter the elastic behaviour of matter was was discussed. Under conditions of elastic deformation a solid acquires a given deformation when a given load is applied, and this deformation disappears completely when the load is removed. If the strain is made too large the behaviour of a solid is no longer purely elastic. Part of the deformation is not recovered on unloading, the specimen is said to acquire a 'permanent set' and to have suffered 'plastic deformation'.

The remainder of this section deals with the definition of some terms that are of general application to non-polymeric solids, but the rest of the chapter, apart from section 3.2, deals with the permanent deformation of crystalline materials only, with the emphasis on metals.

The differences in behaviour of different materials under load have given rise to a number of different terms to describe their behaviour. Materials that need high stresses to produce even small strains, that is, have a high Young's modulus, are termed 'stiff'. An important property possessed by some materials is the ability to absorb large amounts of strain energy when permanently deformed, and to break gradually if loaded sufficiently. Such materials are termed 'tough'. A 'strong' material, on the other hand, is one that shows a high yield stress, and can stand an even higher stress before it fractures.

For a material to be tough it must be able to suffer some plastic deformation. High strength, however, implies a resistance to plastic deformation, and such a material may, in fact, fracture in a 'brittle' manner, in which fracture occurs rapidly and catastrophically after very little plastic deformation.

Two further terms are frequently used to describe the plastic properties of materials, particularly of metals. These are 'malleability', which is the ability to be beaten into a thin sheet without splitting, and 'ductility' which is the ability to be drawn into a wire easily, though the term 'ductile' is often used to describe a material that deforms plastically before fracture.

3.2 The tensile test

One of the most widely used ways of examining the mechanical properties of a material is to subject a suitable sample to a tensile test. In a tensile test a specimen is extended axially, and corresponding values of load and extension are measured. From these, graphs showing the

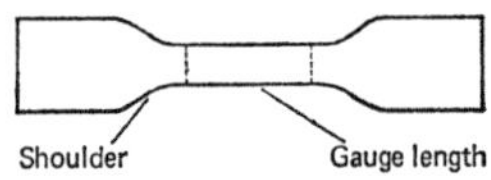

Figure 3.1

relationship between tensile stress and strain can be obtained. To ensure that the deformation is quite axial, care must be taken in the alignment of the specimen and in the way the ends of the specimen are gripped. In order to eliminate complications that might arise from the way the ends of the specimen are gripped (it is not usually convenient to pull directly on the ends of a specimen), the specimen is made much longer than its lateral size, and is often given a cylindrical shape with a principal section as shown in fig. 3.1. The specimen is gripped at the enlarged ends. In these regions stresses will not be pure tensile stresses, but will have values lower than those in the central narrower section. The measured length to be examined, the 'gauge length', is in this narrower section. In this region, well away from the ends, the stress is a uniform pure tensile stress, at least for small deformations. When large plastic deformations occur, transverse stresses may be set up.

Frequently the applied force is measured by means of a 'load cell'. One arrangement is a network of resistance strain gauges mounted on a stiff ring. The load is applied across a diameter of the ring, which is sufficiently stiff so that it only deforms elastically under all the loads encountered. The distortion of the ring changes the resistances of the strain gauges, the magnitude of the change depending on the value of the load. Usually the resistance network is balanced for zero load, and the out-of-balance voltage measured as a function of the load.

Modern testing machines usually deform the specimen at a constant extension rate by moving one end of the specimen at a constant velocity, while the other end remains fixed. The specimen then deforms to maintain this imposed extension rate, and the load cell measures the force needed to maintain this extension rate. A schematic representation of an apparatus to perform a tensile test in this manner is shown in

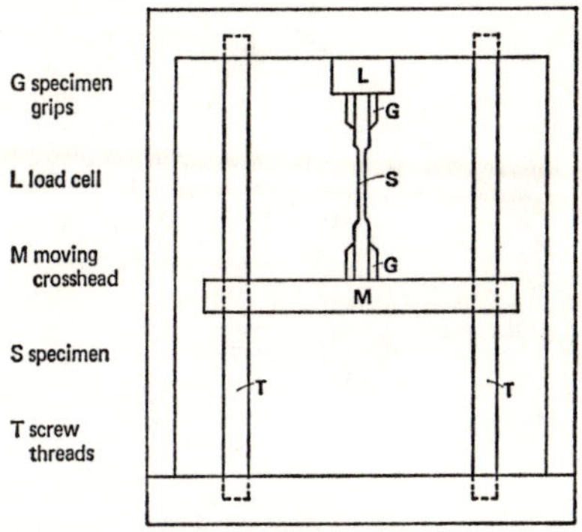

Figure 3.2

fig. 3.2. The apparatus is mounted with the specimen axis vertical. Two accurate screw threads TT are rotated together at a constant rate, and this causes the cross-head M to move at a steady rate, thus forcing the lower grip to move. The resistance of the specimen, which is the load needed to produce the given extension rate, is measured by the load cell L. The motor, electrical equipment and so on are not shown. The apparatus is easily modified to allow the tests to take place at different temperatures and also at different extension rates.

3.3 Features of the tensile test

Many common metals, such as copper and aluminium, are ductile at room temperature. If a tensile test is carried out on a polycrystalline sample of such a metal, and the results obtained presented on a conventional stress, σ_C, against strain, ϵ, graph, the curve obtained has the form shown in fig. 3.3. Over the region OA the graph is a straight line passing through the origin, the behaviour is perfectly elastic and Hooke's Law is obeyed. When the stress exceeds that at A, macroscopic permanent deformation occurs, so that the stress at A is the yield stress, σ_Y, and A is the yield point (neglecting the slight difference between yield point and proportional limit). The exact position of A is often uncertain and sensitive techniques show that, before plastic deformation takes place on a macroscopic scale, that is, while the crystal as a whole deforms elastically, there is usually some very localized plastic deformation. The strain corresponding to A is much less than 1% in general. As

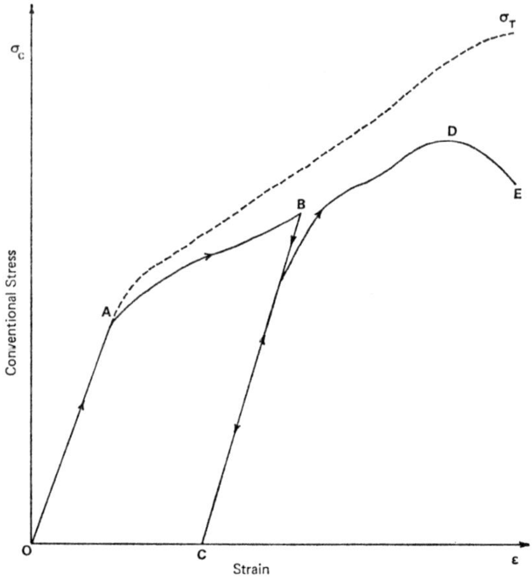

Figure 3.3

soon as plastic deformation starts, the stress-strain curve bends towards the strain axis. However, the curve continues to rise, so that the stress needed to produce further plastic deformation increases with the strain, that is, the material gets progressively harder. This effect is called 'work hardening' or 'strain hardening'. If work hardening were absent, the load that a tensile test piece could sustain would continually decrease as extension proceeded, owing to the decrease in the area of cross-section.

If the test piece is unloaded when the point B is reached, the unloading path is BC, which has approximately the same slope as OA. The elastic deformation is recovered, but the unloaded bar retains a strain OC which is the permanent set. The stress needed to produce some specified plastic strain is referred to as a 'proof stress'. For example, if OC represents a strain of 0·1%, the stress at B is the '0·1% proof stress'.

When the test is continued, the stress rises along OB, but before B is reached the curve bends over towards the strain axis, and then con-

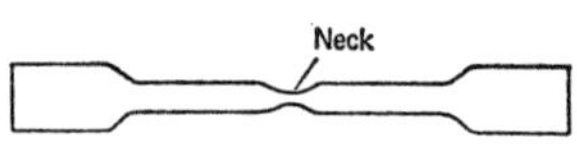

Figure 3.4

tinues as if the unloading had not taken place. The conventional stress continues to rise until D is reached and then starts to fall. The load corresponding to D is the maximum load that the specimen can withstand, and the value of the conventional stress at this point is called the 'ultimate tensile strength' σ_U.

Over the range of deformation represented by OD the extension is homogeneous, that is, on the macroscopic scale the deformation is the same for all cross-sections in the gauge length. At D, however, a 'neck'

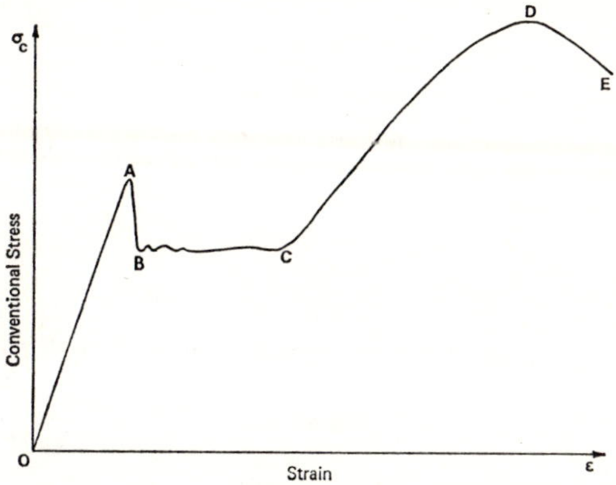

Figure 3.5

forms in the specimen, as shown in fig. 3.4; subsequent plastic extension is restricted to this neck and the load needed to produce further extension falls. The neck gets progressively narrower until fracture occurs at E. 'Necking' is a characteristic of the tensile test, and arises because the work hardening resulting from the extension is not sufficient to offset the 'geometrical softening' that arises due to the decrease in the area of cross-section.

If true stress σ_T is plotted instead of σ_c, there is very little difference in the curves for small strains, but as the strain increases the difference gets bigger. σ_T is, of course, larger than σ_c. When the point of instability, D, is reached, if σ_T is measured in the neck, that is, σ_T = load $\div$ area of cross-section of neck, σ_T continues to increase right up to the point of fracture. The variation of σ_T with ϵ is shown as a dotted line in fig. 3.3.

Several different features are obtained from tests carried out on materials of which annealed mild steel and brasses with 20–30% by weight of zinc are typical. The $\sigma_c - \epsilon$ curve in these cases takes the general form shown in fig. 3.5. Starting at 0, a long elastic range is sharply terminated at A when the stress reaches a value known as the 'upper yield stress'. There is an abrupt partial unloading, and macroscopic plastic deformation begins locally. The regions which have been plastically deformed are called 'Lüders bands'. As the extension of the specimen proceeds, the Lüders bands spread along the whole gauge length, until at C they cover the whole specimen. During this process the value of σ_c oscillates about a relatively constant value, known as the 'lower yield stress'. Once the localized deformation has spread over the whole of the specimen, further deformation is homogeneous throughout the specimen, and σ_c rises as the material work hardens, to give a maximum value at D when a neck occurs, resulting in subsequent fracture at E.

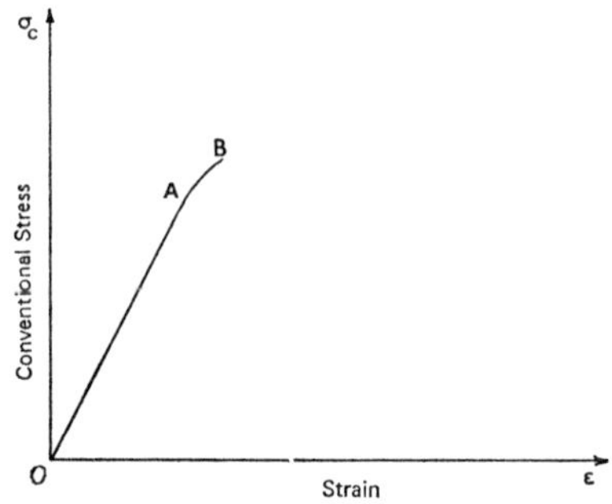

Figure 3.6

Finally, if a brittle material is tested in tension, a $\sigma_c - \epsilon$ curve of the form shown in fig. 3.6 is obtained. The curve shows an elastic region OA, followed by a very small amount of plastic deformation before fracture occurs at B. No neck is formed.

Stress-strain curves for single crystals show the same basic features as tests on polycrystals, though the magnitudes of the details depend on the orientation of the crystal lattice with respect to the tensile axis. The true stress-strain curve for a copper single crystal at room temperature is shown in fig. 3.7.

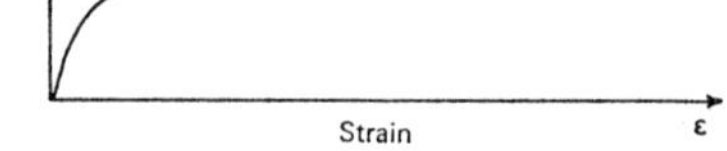

Figure 3.7

3.4 Experimental values

The values obtained experimentally for such quantities as yield stress and ultimate tensile strength depend to some extent on a number of variables, such as purity of the material, grain size, temperature of testing and strain rate. Some values for pure polycrystalline materials in the form of wires tested at about room temperature are given in Table 3.1. It will be noted that different materials have very different values for these properties. This is not the case with Young's modulus, where quite different materials have values of the same order of magnitude.

Table 3.1

Material	Yield stress, σ_Y N m^{-2}	Ultimate tensile strength, σ_U N m^{-2}
Copper	40×10^6	160×10^6
Aluminium	26×10^6	60×10^6
Mild steel	$0{\cdot}4\times10^9$ (upper yield point) $0{\cdot}3\times10^9$ (lower yield point)	460×10^6

Values of the yield stress for single crystals are usually much smaller than for polycrystals at the same temperature.

3.5 Plastic deformation by glide

Three distinct processes of plastic deformation in crystals are usually recognized. These are glide (or slip), twinning and kinking. All three processes are simple shears.

When a crystal deforms by mechanical twinning, the deformed regions suffer a simple shear on a fixed crystallographic plane, the 'twinning plane', in a fixed crystallographic direction, the 'twinning direction'. This shear is homogeneous throughout the whole of the twinned region. The orientation of the twinned lattice differs from that of the 'parent' or 'matrix' lattice. Either (*a*) the twin lattice is a mirror image of the matrix lattice with the plane of twinning as the mirror, or (*b*) the twin lattice is obtained by rotating the matrix through 180° around the direction of twinning. In the simplest cases both relationships are fulfilled at the same time.

On the basic particle scale what happens is that successive planes of particles parallel to the twinning plane slide over each other in the twinning direction. Each plane moves over the one below it a fixed distance that is a fraction of the particle spacing in the twinning direction. As a result of this small amount of shear the particles in the deformed region arrive at a configuration that is identical (or becomes identical after small additional movements known as 'shuffles') with the original lattice, but differs in orientation.

Fig. 3.8a illustrates mechanical twinning in a simple lattice. It shows a section of the lattice perpendicular to the plane of twinning and containing the twinning direction. The small circles represent the positions of the particles and the arrows the shearing forces producing twinning. The plane of twinning is perpendicular to the plane of the paper and is indicated by the horizontal line TT. The full lines show the lattice in both parent and twin regions after deformation; the dotted lines show the lattice above the twinning plane before twinning. The dotted arcs show the movement of the particles.

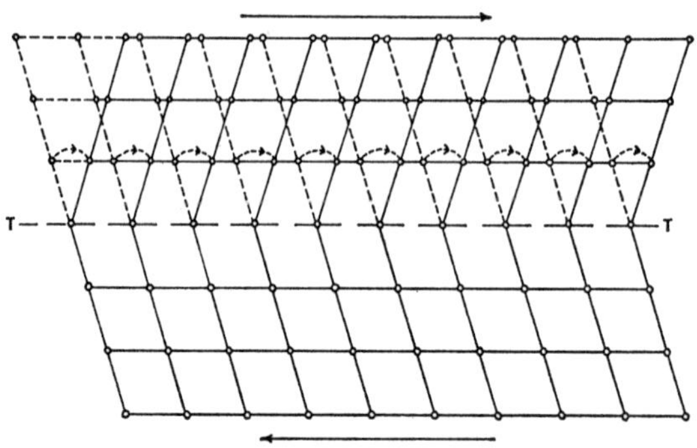

Figure 3.8a

When kinking occurs the crystal breaks up into regions of equal thickness, called 'glide packets', which, sliding upon each other, swivel over into another position. In this process they turn around 'hinges' situated in the 'plane of kinking'. In their new positions these glide packets, but not, in general, the lattice, are mirror images of the corresponding (unoperative) packets in the undeformed part of the crystal, the plane of kinking being the mirror. The glide packets are usually too thin to be microscopically observable.

Kinking is macroscopically homogeneous and the lattice orientation in the deformed parts is different from that in the undeformed part. It differs from twinning in that the plane of kinking and the amount of shear are not, in general, crystallographically fixed.

A schematic representation of kinking is given in fig. 3.8b. It shows a section of the sample perpendicular to the plane of kinking and containing the direction of kinking. The arrows indicate the shearing forces and the dotted line KK represents the plane of kinking. Each glide packet contains several, usually many, lattice glide planes. Note that neither molecules nor lattice glide planes are shown here. The dotted lines show the positions of the active glide planes before kinking and the full lines the positions of the glide packets (both operative and inoperative) after kinking. The double lines shown indicate a certain lattice direction. During kinking a lattice direction remains fixed with respect to the lattice of the glide packets and rotates with it. Hence lattice directions in both the kinked region and the matrix, in general, are not symmetrical with respect to the plane of kinking.

In most crystals glide is the dominant process giving rise to plastic deformation and will therefore be considered in greater detail than twinning or kinking.

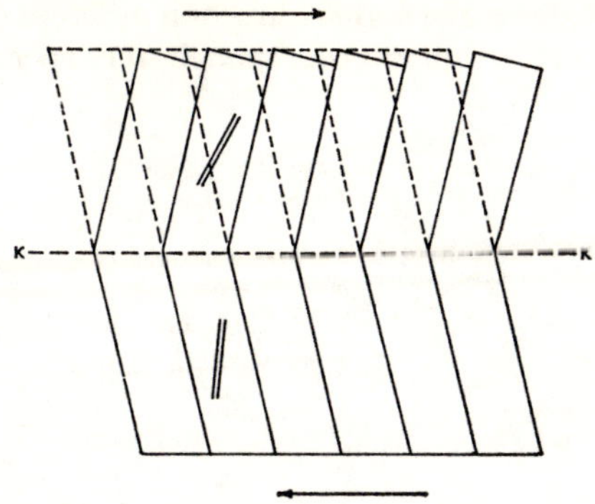

Figure 3.8b

When a crystal with polished surfaces is deformed plastically, and then examined under a low-power optical microscope, the surfaces are, in general, found to be crossed by a number of fine dark lines known as 'slip bands' or 'glide bands'.* They arise from the glide process in which adjacent blocks of particles slide over one another. Because of the strong periodic nature of the interparticle forces in solids, the blocks will slide over each other by one or more complete particle spacings. This process results in a small 'slip step' being produced on the surface

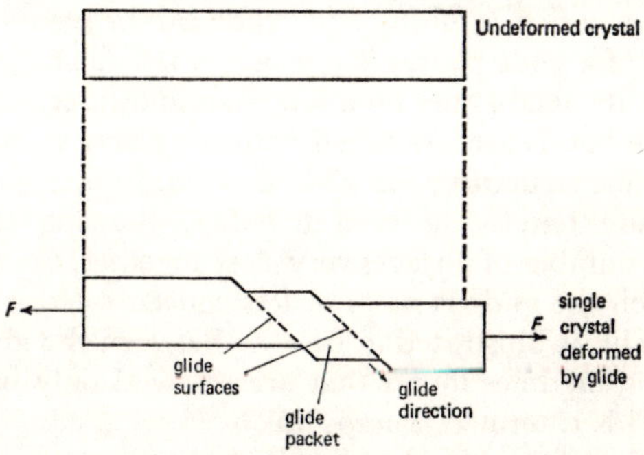

Figure 3.8c

of the crystal. Under a low-power microscope this step appears as a fine line. The surface on which glide has taken place is called the 'glide surface'. In many cases this surface is planar, in which case it is referred to as the 'glide plane'. Experiment shows that the relative displacement

* The use of the terms glide or slip seems to be a matter of personal taste.

of two adjacent blocks of particles is always in a particular crystallographic direction, usually that along which the particles are most closely packed. This direction is the 'glide direction'. The glide plane is less characteristic; for a given glide direction different glide planes may operate at different temperatures, though the operative planes are

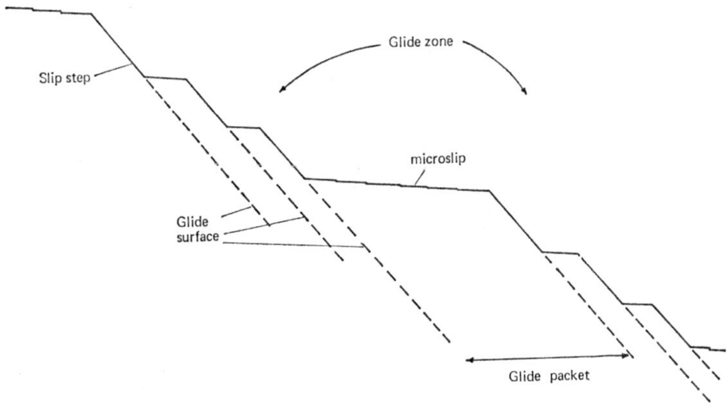

Figure 3.9

usually crystallographic planes of high particle density. The parts of the crystal between glide surfaces are not deformed plastically, but suffer only elastic deformation, and are called 'glide packets'. Fig. 3.8c shows schematically the glide process in a single crystal deformed in tension. The extension resulting from glide can be seen.

The width of a glide packet is typically a few micrometres and each slab slips over its neighbours by a few thousand interparticle spacings.

When glide bands are examined with the electron microscope they often show a fine structure; the glide band is, in fact, found to consist of several glide bands clustered together, showing that glide has occurred on a number of surfaces very close together. For this reason the region in which the glide is more or less concentrated is often called a 'glide zone'. This is illustrated in fig. 3.9. Between the glide zones finer markings are sometimes found that are resolved only by the electron microscope. This is termed 'micro-slip'.

Thus, plastic deformation by glide is not homogeneous on the microscopic scale, since gliding is restricted to a few widely spaced groups of particle planes. This is in contrast with elastic deformation in which all planes of particles are affected.

Fig. 3.8c shows clearly that when a single crystal extends by glide, blocks of particles slide over their neighbours in particular direction. As a result of this, the two ends of the crystal are displaced laterally with respect to each other. Displacement of the ends of the specimen in this way is usually prevented in testing, so that bending of the crystal in the region near the grips occurs, as shown schematically in fig. 3.10, where the lines crossing the specimen represent glide bands. This is another important reason for using long specimens. In addition to bending near the ends of the specimen, the glide packets in the gauge length rotate

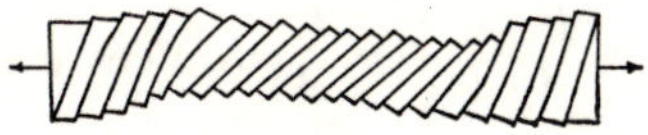

Figure 3.10

about an axis in the glide plane, in such a sense that the angle between the glide direction and the tensile axis is reduced. This rotation of the glide plane and direction also means that, after sufficient extension, the stress on another slip system may become sufficient for glide to start on that, as well as continuing on the original system.

3.6 The laws of glide

One fundamental experimental observation is that, except at high pressures, the application of a uniform hydrostatic pressure to a solid does not lead to plastic deformation. Application of the methods of section 2.7 show that, if a solid is subjected to a uniform hydrostatic pressure, the shear stress on any plane in the solid is zero. From this it may be concluded that plastic deformation is governed by the shear stresses, and the important ones will be the components of shear stress acting in the operative glide planes, resolved along the glide direction. Thus, one

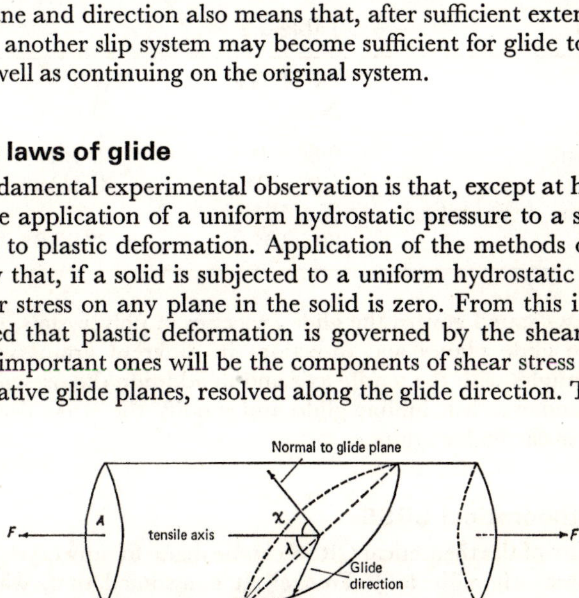

Figure 3.11

basic law of glide is that the plastic strain depends only on the shear stress in the glide plane, resolved parallel to the glide direction. Another important experimental law is that appreciable plastic glide starts when the shear stress on the glide plane, resolved in the glide direction, reaches a fairly well-defined value called the 'critical resolved shear stress' (CRSS). This is the yield stress law of E. Schmidt.

It is a simple matter to calculate the CRSS for a single crystal. Consider the cylindrical single crystal shown in fig. 3.11. It has an area of cross-section A and forces F are applied axially at each end. The normal to the active slip plane makes an angle χ, and the glide direction an angle λ, with the tensile axis. The component of F parallel to the operative glide direction is $F \cos \lambda$ and the area of the glide plane is $A/\cos \chi$. Thus, the value of the resolved shear stress τ_Y is

$$\tau_Y = \frac{F \cos \lambda}{A/\cos \chi} = \frac{F}{A} \cos \lambda \cos \chi$$

Macroscopic plastic deformation is observed when this quantity reaches a critical value for a given material at a given temperature.

Values of the CRSS at room temperature for some single crystals are given in Table 3.2.

Table 3.2

Material	CRSS (τ_Y) $N\,m^{-2}$	$G/6$ $N\,m^{-2}$
Aluminium	$0{\cdot}86 \times 10^6$	25×10^9
Copper	$1{\cdot}0 \times 10^6$	46×10^9
Aluminium + 6% silver	85×10^6	
Iron	30×10^6	77×10^9

The laws stated govern the glide processes in polycrystals, but there the active glide planes and directions in different grains will make different angles with the tensile axis and in addition the presence of the grain boundaries will inhibit glide and modify the stress distribution acting in a particular grain.

3.7 The theoretical CRSS

An estimate of the theoretical CRSS can be made for any crystal using a method due originally to J. Frenkel. It is assumed that, when glide occurs, all the particles in one block of the crystal slide simultaneously over those in an adjacent block. Consider a simple crystal in which the spacing between planes of particles is a and that between particles in any row in a plane is b. It is assumed that the planes chosen are possible glide planes, and that the rows of particles in these planes are parallel to the glide direction. Fig. 3.12a shows two such rows (in adjacent planes) when the crystal is not subject to a stress. Let τ be the shear stress necessary to displace (in the direction of one of the rows of particles) the whole of the upper plane of particles a distance x with respect to the lower plane of particles, as in fig. 3.12b. τ is zero when $x = 0, b, 2b, \ldots$ since the particles are then at the normal lattice sites. Also, when $x = b/2$, the displaced particles are in metastable positions and τ is again zero. Thus τ must be a periodic function of x with a period b. Frenkel assumed that τ could be represented by

$$\tau = k \sin\left(\frac{2\pi x}{b}\right)$$

The value of k is obtained from the condition that Hooke's Law is obeyed when the displacement x is small. Under these conditions

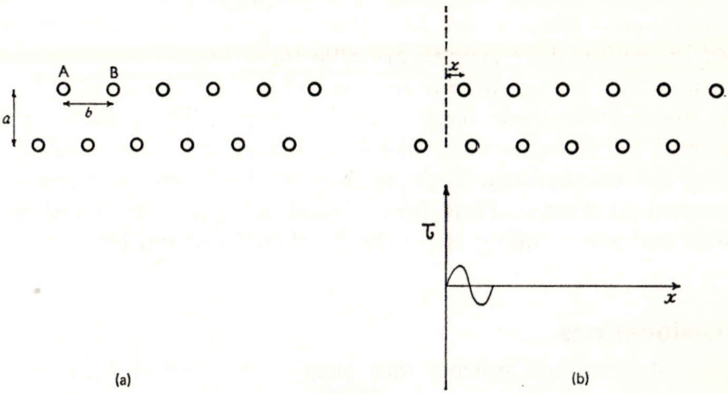

Figure 3.12

$\sin(2\pi x/b)$ is closely equal to $2\pi x/b$. Also, the shear strain is x/a, so that Hooke's Law may be written $\tau = Gx/a$, where G is the rigidity modulus. Equating the two expressions for τ when x is small gives

$$\frac{Gx}{a} = k\frac{2\pi x}{b}$$

or,

$$k = \frac{Gb}{2\pi a}$$

and hence

$$\tau = \frac{Gb}{2\pi a}\sin\left(\frac{2\pi x}{b}\right)$$

The CRSS is τ_m, the maximum value of τ, since, when this value is reached the deformation becomes unstable. The maximum value of the sine function is unity and hence

$$\tau_m = \frac{Gb}{2\pi a}$$

This equation predicts that glide will occur in the direction for which b is a minimum and a is a maximum, that is, it predicts glide in close-packed directions on close-packed planes. For many crystals $a \simeq b$ and the theoretical CRSS on this model is thus about $G/2\pi$ or $G/6$. Table 3.2 shows that this is about 10^4 or 10^5 times greater than the observed values for annealed crystals. Refinements of the above calculation using more realistic laws of force between molecules only reduce the predicted value of τ_m to about $G/30$, which is still much larger than the observed values.

Exceptions to the above observations are 'whisker' crystals. These are very thin single crystals having diameters of about 10^{-6} m. According to the theory of their growth, they should contain a single screw dislocation (see section 3.8) along their length, but no other observable imperfections. Observations show that, in general, this is true. When tested in tension these crystals are able to withstand very high elastic strains ($\sim$2%) before there is any measurable deviation from Hooke's Law. Beyond this strain large deviations occur. The maximum stress supported by these crystals, which occurs in the elastic region, but beyond the proportional limit, is close to the lowest estimate of the theoretical yield stress. Thus these whiskers are probably almost perfect crystals and are breaking at nearly the theoretical yield stress.

3.8 Dislocations

Surface observations indicate that plastic deformation does occur by glide, but the calculation of the CRSS in a single crystal shows that glide is not achieved by a simultaneous movement of all the particles on the glide plane. The only alternative is that glide occurs by the consecutive slipping of particles over each other. In terms of chemical bonds between particles, this would mean the breaking of bonds in turn rather than all at once.

If glide occurs over a complete glide plane in a single crystal, the only consequence is that there is a shear displacement of one block of particles, relative to an adjacent block, by a whole number of particle spacings. However, if glide takes place over part of the glide plane only, a line can be drawn on the glide plane separating those parts where glide has occurred and those where it has not, as in fig. 3.13. This line is termed a 'dislocation line'. The boundary between the slipped and unslipped regions on a glide plane will mark a region of particle misfit; the particle arrangement in this region of misfit is a 'dislocation'. When a dislocation passes through a region of a crystal, blocks of particles on either side of the glide plane undergo a certain relative displacement, which can be represented by a vector called the Burgers vector $\boldsymbol{b}$ of magnitude b. The Burgers vector is associated with the whole dislocation line and must be constant along it. This idea that glide is the result of the movement of dislocations through a crystal was put forward independently by G. I. Taylor, E. Orowan and M. Polanyi in 1934.

By its definition the movement of a dislocation produces glide. It is now necessary to show that a dislocation line in a crystal possesses the essential property of moving under a very small stress. Before this can be done, the particle arrangement in the region of the dislocation line must be determined, at least qualitatively.

If the Burgers vector is everywhere perpendicular to the dislocation line, the dislocation is called an 'edge' dislocation; if $\boldsymbol{b}$ is everywhere parallel to the dislocation line, it is a 'screw' dislocation. In general,

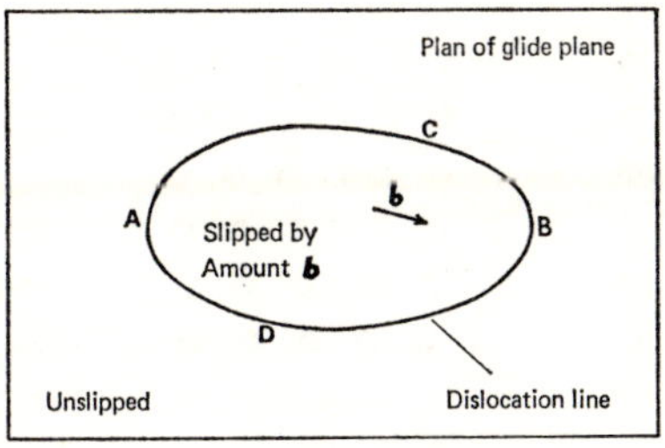

Figure 3.13

dislocations are neither pure edge nor pure screw, and are termed 'mixed' dislocations. In fig. 3.13 the dislocation is pure edge at A and B, pure screw at C and D and mixed at all other points.

An edge dislocation is found wherever glide takes place over part of a glide plane with the Burgers vector everywhere normal to the line separating the slipped and unslipped regions. In fig. 3.14 glide has taken place over the part ABCD of the glide plane ABEF, the Burgers vector being $\boldsymbol{b}$. CD is then the edge dislocation line.

An edge dislocation in the simple cubic lattice is shown in fig. 3.15.

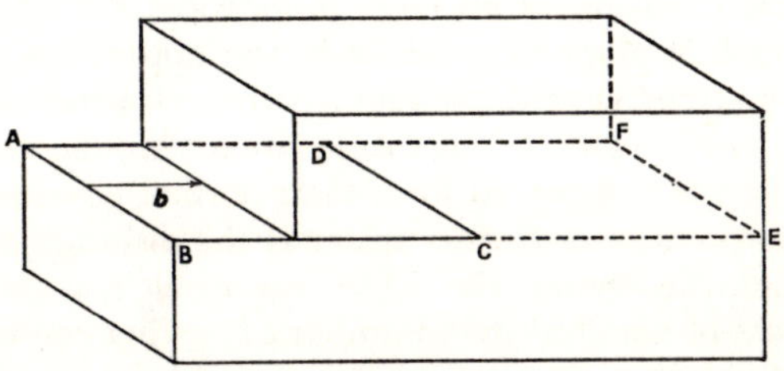

Figure 3.14

Fig. 3.15a shows the particle arrangement in the perfect crystal. The lattice spacing is b and AB is the plane on which glide is going to occur. Since the crystal lattice provides a periodic field of force, particles will glide from one equilibrium position to the next under the influence of a stress, so that the Burgers vector will be parallel to the rows of particles, and its magnitude will be a lattice spacing. Let glide occur by an amount b over the part of the glide plane AC, but not over CB, as in fig. 3.15b. Then C indicates the dislocation. If all particle planes parallel to the plane of the paper have suffered the same deformation, the dislocation line extends through C in a direction normal to the plane of the paper. Fig. 3.15b shows that the introduction of an edge dislocation into a crystal is equivalent to making a cut in the perfect crystal, normal to the glide plane, and inserting an extra half plane of particles CD. This

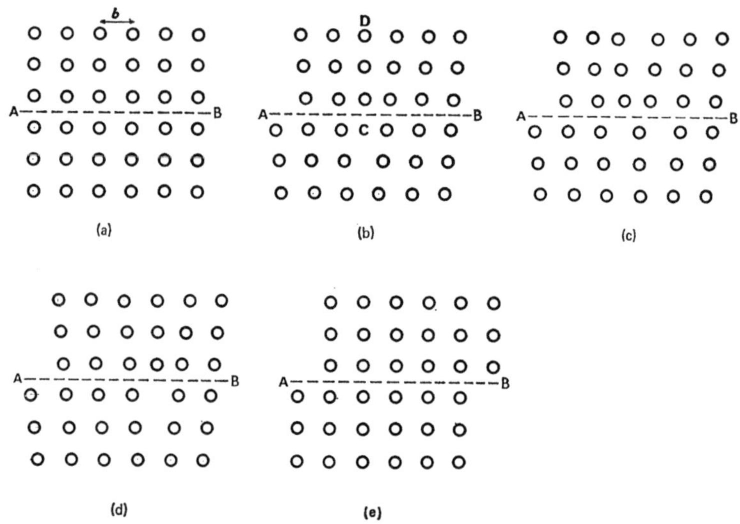

Figure 3.15

is the distinctive feature of an edge dislocation. An edge dislocation with the extra half plane of particles in the upper half of the sample is termed a 'positive' edge dislocation. If the extra half plane is in the lower half, it is a 'negative 'edge dislocation. Around the dislocation line the particles are displaced from their normal positions, and hence dislocations are centres of elastic strain. If the dislocation passes completely through the crystal (fig. 3.15e) the result is a shearing of two adjacent blocks of material by an amount b, giving rise to two surface steps, each of height b, and leaving the crystal lattice undistorted.

For a screw dislocation, lattice shearing takes place parallel to the dislocation line, so that in a simple cubic lattice, where glide by the movement of a screw dislocation has taken place over part of a glide plane, the situation is as shown in fig. 3.16. Glide has occurred over the plane ABCD, the Burgers vector being $\boldsymbol{b}$, DC is the screw dislocation line. Starting at E and following a path which passes once completely round the screw dislocation, such as EFGHK, it can be seen that such a path is a helix of pitch b. For this reason the dislocation is referred to as a screw dislocation. Once a crystal contains a screw dislocation all lattice planes are really parts of a helix, but away from the dislocation line the displacement of particles from their normal positions is so small that the layers of particles are effectively planes.

3.9 The movement of a dislocation in an otherwise perfect crystal

Consider the edge dislocation line through C in fig. 3.15b. Particles on the glide plane AB that are sufficiently far ahead of, or behind, C are in their normal positions, but those close to the dislocation are displaced from these positions. In the region of C, particles above the

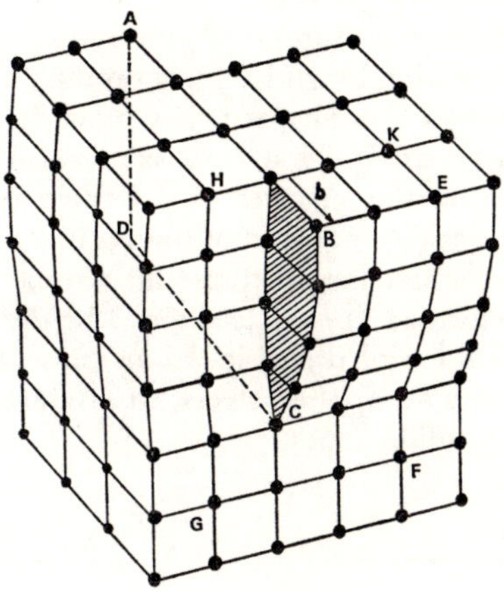

Figure 3.16

glide plane are closer together than normal, and those below are farther apart. The dislocation is under symmetrical forces with respect to movement on the glide plane. Hence, it is in an equilibrium position, and needs a very small force applied to it to move it slightly to left or right. In passing from one stable equilibrium position like this (fig. 3.15b) to an adjacent one (fig. 3.15d) the dislocation passes through another symmetrical arrangement, as in fig. 3.15c, where the forces acting on the dislocation are in equilibrium. Thus, the force needed to move the dislocation is that needed to move it from the arrangement of fig. 3.15b to that of fig. 3.15c. Calculation of the exact force needed to move the dislocation is difficult, but clearly it is small. Similar considerations show that the movement of the screw dislocation through a crystal can occur under a very small stress.

Detailed calculations show that the stress needed to move a dislocation through an otherwise perfect lattice is, in fact, less than the observed CRSS. Also, if this were the governing mechanism, the CRSS would have the same value for all specimens of a given material, whereas in practice the CRSS depends to some extent on the way the crystal has been prepared. This problem is discussed further in section 3.11.

The movement of a dislocation in its glide plane, defined as the plane containing the dislocation line and its Burgers vector, is termed glide. Pure screw dislocations can thus glide on any allowed glide plane. Edge dislocations can glide on their own glide planes only. They can, however, move in directions at right angles to $\boldsymbol{b}$ if particles are added to, or removed from, the extra half plane. This movement is called 'climb'.

3.10 Dislocation multiplication

Dislocations have been shown to satisfy the requirements of a mechanism for plastic deformation by glide. They produce a permanent shearing of adjacent parts of a crystal when they move on a glide plane, and this movement can occur under a very small stress. To produce the observed glide steps on the surfaces of deformed crystals, very large numbers of dislocations (several tens of thousands) must move on each glide plane. This is much larger than the numbers present in such regions in the undeformed crystal (section 3.12 discusses 'seeing' dislocations), and thus some of the dislocations present initially must be able to multiply under an applied stress, to produce large amounts of glide in relatively few glide zones.

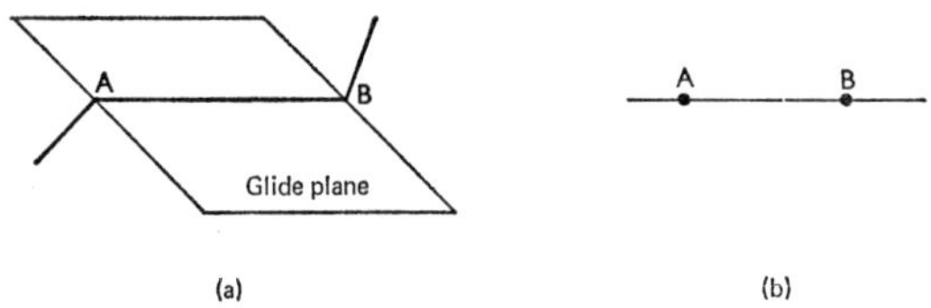

Figure 3.17

One important mechanism that operates in the interior of some crystals is the 'Frank-Read source'. A segment of a dislocation line lying in a glide plane is imagined pinned at two points, as at A and B in fig. 3.17. The pinning may be due, for example, to the remainder of the dislocation line not lying in the glide plane and not being pure screw, as in fig. 3.17a, or to impurities or inclusions preventing the dislocation from moving at A and B, as in fig. 3.17b. When a stress is applied, only the segment AB of the location line is free to move. In plan the glide plane appears as in fig. 3.18. When no stress is applied the dislocation line has position 1. It is held back at A and B when a stress is applied, and thus it bows out to position 2. Continued application of a sufficient stress causes it to take up successive positions 3, 4, 5. When the loop reaches position 6 it intersects itself, leaving a complete dislocation loop which continues to expand, and a short segment lying between

A and B which can then repeat the process described. In this way large amounts of glide can be obtained from a single initial dislocation line.

It can be shown that the maximum stress needed to operate the source corresponds to the situation when the dislocation loop is semicircular, and is approximately $Gb \div AB$, where G is the rigidity modulus of the material (assumed isotropic).

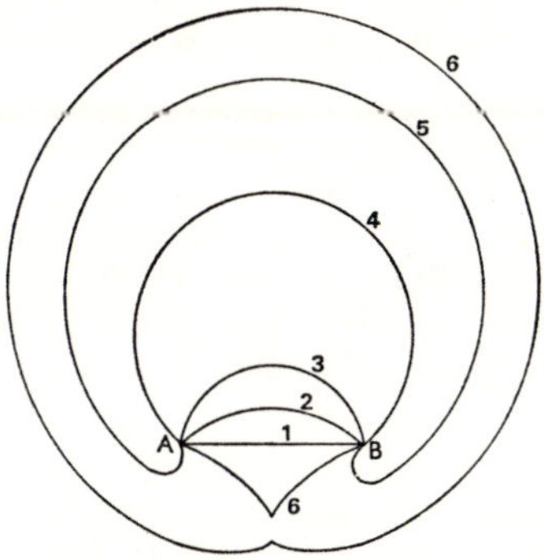

Figure 3.18

3.11 The CRSS and work hardening

Macroscopic plastic deformation first occurs when the resolved shear stress reaches a critical value. At the basic particle level, this means that a large number of dislocations starts to move in each glide zone. One dislocation moving completely through a crystal produces a shear displacement of b, that is about $0{\cdot}5 \times 10^{-9}$ m. Thus, if N dislocations of Burger vector $\boldsymbol{b}$ move through a cube of crystal of side 0·01 m, on a glide plane parallel to one of the cube faces, as in fig. 3.19, the shear of the top surface relative to the base is Nb. If this shear strain is 0·1%, this gives a value of N of 2×10^4, indicating that very many dislocations

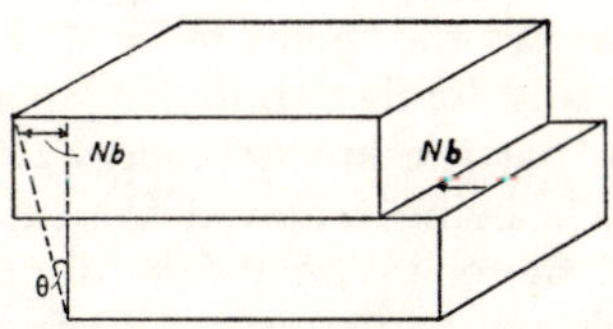

Figure 3.19

must pass completely through the crystal to give this small strain. If they do not pass right through the crystal, many more are needed. Thus the CRSS is the stress needed to cause the movement of a large number of dislocations.

Even well-annealed pure crystals usually contain a large number of 'grown-in' dislocations (often up to 10^{10} m^{-2}). The dislocations responsible for glide must move in the presence of these dislocations, and may have to cut through them if they cross the active glide plane. The CRSS is determined by the force needed to produce this movement. Thus, without a detailed consideration of particular mechanisms, it can

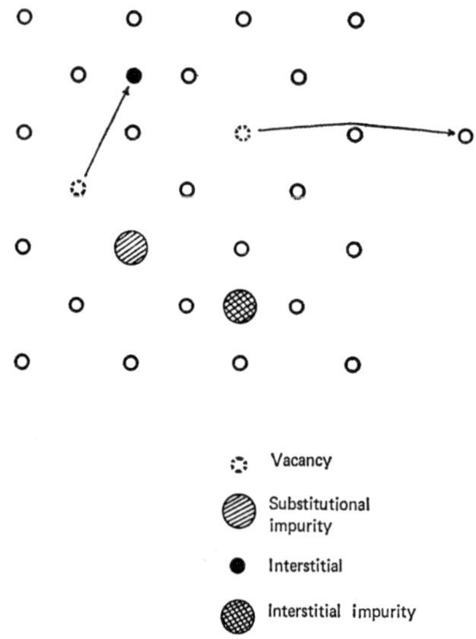

Figure 3.20

be seen that the CRSS is really a measure of the forces that dislocations exert on each other, rather than the stress to move a single dislocation through an otherwise perfect lattice, though the smallness of this latter stress is reflected in the small value of the CRSS.

The CRSS can also be influenced by interactions between dislocations and other defects. One important group of defects that can interact with dislocations comprises the 'point defects'. Examples of these are vacancies, interstitials and foreign (impurity) particles (see fig. 3.20).

The fluctuations in particle kinetic energy that occur in a crystal may result in a particle being removed from its normal lattice site, thus creating a vacant site or vacancy. The displaced particle may be considered as being ejected to the crystal surface, or to a dislocation line or grain boundary, leaving an isolated vacancy known as a Schottky defect. Alternatively the displaced particle may be forced into a position between normal lattice sites. It is then termed an interstitial. The interstitial and vacant lattice site together constitute a Frenkel defect.

Vacant lattice sites (giving Schottky and Frenkel defects) are, in fact, continually being created and destroyed in a crystal at all temperatures, their equilibrium concentration increasing rapidly with temperature. However, the energy of formation of the two types of defect is different in most materials, and then one type (usually the Schottky defect) predominates.

In addition to Schottky and Frenkel defects, impurity particles may be present, either occupying normal lattice sites, that is, filling vacancies, in which case they are substitional impurities, or occupying interstitial sites, when they are interstitial impurities.

These defects can reduce the elastic energy of the material by taking up positions close to the dislocation, where they partly cancel the lattice distortion produced by the dislocation. For example, a large interstitial impurity particle would reduce the distortion by moving to the expanded region round an edge dislocation. To make the dislocation move in the presence of point defects, a stress must be applied sufficient to overcome the elastic interaction between the dislocation and the point defect.

Once plastic deformation has started in a material, an increase in stress is needed to produce further deformation, that is, as deformation proceeds, the movement of dislocations on their glide planes becomes more and more difficult. This hardening may arise from elastic interactions between dislocations, since there is a strain field round a dislocation, but it may also arise, in part perhaps, from elastic interactions with other defects in the crystal lattice. These may be point defects or surface or volume defects.

Grain boundaries, and also the crystal surface, constitute two-dimensional defects where the particle arrangement is not that of the perfect material. Grain boundaries in particular, are regions of considerable local lattice distortion and hence, like dislocation lines, are preferred regions for the segregation of impurity particles which distort the lattice. Also, grain boundaries are often able to hold up dislocations because of the change of orientation of the glide plane that occurs on crossing such a boundary. For these reasons polycrystalline samples of a given material are harder than single crystals.

Thus, the interactions between dislocations, and between dislocations and other defects, can determine the stress at which yielding first starts, and also the rate of work hardening.

Materials are usually softened by annealing, that is, heating followed by slow cooling. When a deformed crystal is heated, dislocations are able to glide under the locked-in stresses, and pairs of dislocations of opposite sign annihilate each other. Also, vacancies are able to diffuse more rapidly through the lattice. They are attracted to dislocations, and thus enable the latter to move by climb. By glide and climb the dislocations move into positions relative to each other in which the lattice distortions are reduced, thus reducing the magnitude of the strain in the lattice, and making it easier for dislocations to move through the crystal.

3.12 The direct observation of dislocations

At the present time the only device available for observing dislocations directly, in the sense of seeing the particle arrangement around a dislocation line, is the field ion microscope. In this the specimen is in the form of a fine needle with a hemispherical tip. The tip is kept at a positive potential and around it a high electrostatic field is maintained. The surrounding very low-pressure gas is ionized very close to the tip (where the electric field is most intense), and the ions produced are repelled radially on to a fluorescent screen. The ion current from any region of the specimen tip depends on the local electric field strength, and this is strongly dependent on the particle arrangement in the surface, which thus shows up in the contrast variations in the image on the fluorescent screen.

Other methods of observing dislocations are less direct. For some materials etchants are available that produce etch pits exclusively at the sites where dislocation lines cut the crystal surface. The mechanism giving rise to this preferential etching is not yet clearly understood, and in many cases it is essential that certain impurities, which segregate round the dislocation lines, should be present in the crystal.

Dislocations can also be revealed by 'decoration', in which a second phase is caused to separate preferentially on the dislocations. This allows the three-dimensional arrangement of dislocations to be studied. The decoration process usually involves heating the specimen, so that dislocation arrangements after annealing are revealed, though, in the case of the silver halides, photolytic silver particles produced through exposure to light decorate the dislocations without the need for heating.

A method applicable to transparent isotropic crystals uses the fact that the strain around a dislocation line causes the crystal to become birefringent.

Very thin samples (a few thousand angstroms thick) of many materials can be examined by transmission electron microscopy. The distortion of the lattice around a dislocation line causes changes in the intensity of the diffracted beams from these regions, producing contrast changes in the image. This is a very widely used technique. Similar effects are produced when crystals are examined using X-rays. Dislocations produce contrast variations in the X-ray diffraction pattern.

3.13 Example

In a given crystal there is a uniform density of edge dislocations ρ, with Burgers vector $\boldsymbol{b}$. If the dislocations move with an average velocity v, find an expression for the rate of shear.

Construct a square ABCD (see fig. 3.21) with its plane normal to the glide plane and sides AB, DC of unit length and parallel to $\boldsymbol{b}$. The number of dislocation lines crossing this plane at right angles is ρ. When the dislocations move, since they travel with the same average velocity, the number of dislocations cutting ABCD at any instant remains ρ. When ρ dislocations pass completely through ABCD it is sheared to the shape EFCD. Each dislocation produces a small step in the traces DE

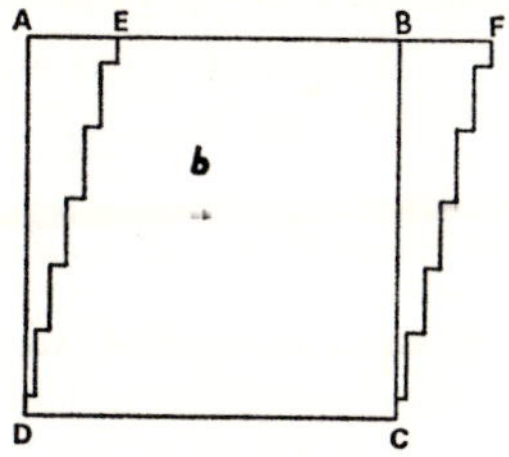

Figure 3.21

and CF so that AE $= \rho b$. Since DA is unity, the macroscopic shear strain is ρb. The time taken for these dislocations to move across ABCD is $1/v$ and hence the rate of shear is $\dfrac{\rho b}{1/v} = \rho v b$

Problems

1. An ideal elastic-plastic solid is one that deforms elastically until the yield point is reached, and thereafter deforms plastically under a constant stress, the yield stress. One such material is isotropic with Young's modulus 80×10^9 N m^{-2} and yield stress 120×10^3 kg m^{-2}. A right circular cylinder of this material, initially 0·2 m long and of radius $1{\cdot}0 \times 10^{-3}$ m, is extended axially to give a total extension of $6{\cdot}0 \times 10^{-3}$ m and is then unloaded. Calculate (*a*) the total work done in producing the final extension, (*b*) the elastic energy recovered, (*c*) the permanent set.

2. An edge dislocation of Burgers vector $\boldsymbol{b}$ is in a region of a solid where the shear stress resolved in the glide plane and in the direction of $\boldsymbol{b}$ is τ. Show that the force F acting on unit length of dislocation line due to the applied stress is $F = \tau b$.

3. When a solid deforms plastically it does so without change of volume. Show that under these conditions Poisson's ratio is equal to 1/2.

4. When the results of a tensile test are plotted on a true stress-strain graph, show that the point at which necking starts is the point of contact of the tangent to the curve that passes through the point on the strain axis where the strain equals -1. (This is Considère's construction.)

5. The true stress-strain relation of an isotropic ductile solid is given by

$$\text{true stress} = 30 \times 10^6 \sqrt{\epsilon}\ \text{kg m}^{-2}$$

where ϵ is the axial strain. Calculate the true stress and strain at which necking begins and hence calculate the ultimate tensile strength.

6. A thin sheet of a crystalline solid is bent to form part of the wall of a right circular cylinder with a mean radius of curvature r by introducing into it a uniform density of edge dislocations. Each dislocation has a Burgers vector $\boldsymbol{b}$ parallel to the neutral plane of bending in the sheet. Show that in the absence of long-range stresses, the density of dislocations needed is $1/rb$.

Further reading

Dekker, A. J., *Solid State Physics*, Prentice-Hall, 1957, ch. 7.

CRANFIELD INSTITUTE OF TECHNOLOGY
LIBRARY

4
fracture

4.1 Introduction

Fractures occurring in tension can, in general, be classed under two headings: those that are preceded by a significant amount of plastic deformation and are referred to as 'ductile' fractures, and those which occur after little or no plastic deformation and are termed 'brittle' fractures. All crystalline materials fracture in a ductile manner if the temperature is high enough, and in a brittle manner at sufficiently low temperatures.

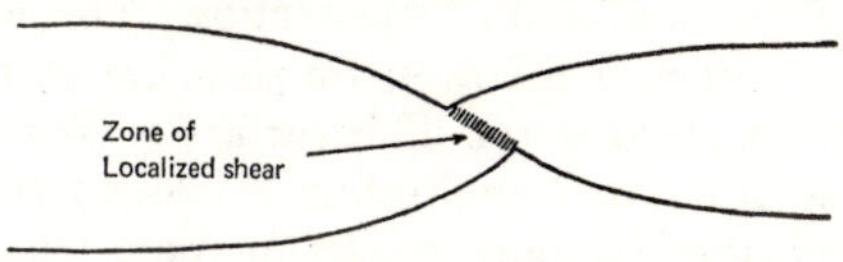

Figure 4.1

4.2 Ductile fracture

Pure single crystals of ductile materials usually fracture in tension in one of two ways. Either (a) the neck develops so that the crystal tapers down to a chisel edge, known as 'rupture' if the reduction in cross-sectional area is 100%, or (b) slip is concentrated in a narrow zone making an angle of about 45° with the tensil axis, which leads to 'slipping off', or 'shearing', giving a fracture as shown in fig. 4.1. Rupture is associated with materials of high purity. Shear is more commonly observed in alloys and also in pure metals deformed at low temperatures.

Slipping-off is observed in many pure polycrystalline materials. In materials of nominal purity, and also in single crystals of low purity, necking is associated with the formation of small holes and cracks in the centre of the neck. These are produced during deformation at inclusions and possibly by dislocation interactions. In the case of cylindrical specimens the formation of the neck is accompanied by the setting up of a system of triaxial stresses at the centre of the neck. Fig. 4.2 shows that,

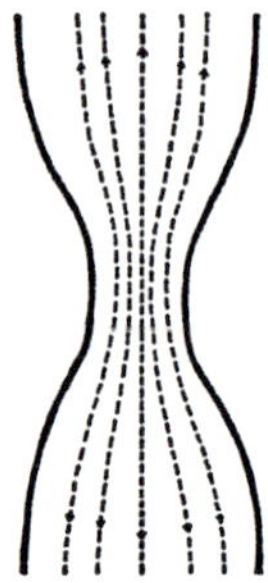

Figure 4.2

as a result of necking, the forces between the adjacent sections must follow the trajectories shown, and thus will have components normal to the specimen axis. This triaxial stress system can be considered as a hydrostatic stress plus a longitudinal stress. The former does not result in plastic deformation and so the material is effectively hardened.

Any holes that are formed in the neck are able to grow transversely more rapidly under the transverse tensions than in the tensile direction and hence are able to coalesce, leading to an internal fracture surface at the centre of the minimum cross-section. The formation of this internal fracture surface is followed by shearing or slipping-off on a surface of maximum shear stress. This surface is conical, giving rise to the 'cup and cone' fracture. Longitudinal sections through the cup and cone fracture typical of (*a*) pure metals at room temperature and (*b*) many common alloys are shown in fig. 4.3.

In both single and polycrystals the path of the parting leading to fracture is determined by macroscopic constraints in the region of the neck. The neck gets progressively narrower, but the remainder of the specimen is not deforming, and remains of constant section. These constraints are influenced by the geometry of the specimen at the start of necking.

4.3 Brittle fracture

Single crystals showing brittle fracture often part on a crystallographic plane of high molecular density. If the fracture occurs after little or no plastic deformation, as in the case of many minerals and some metals, such as bismuth at room temperature, the fracture obeys, in general,

Sohncke's Law, namely, that cleavage occurs when the normal component of the stress acting across the cleavage plane reaches a critical value. Thus, brittle fracture is, in effect, a pulling apart of two parts of the crystal. The tensile stress needed to produce brittle fracture is called the 'brittle strength' of the material.

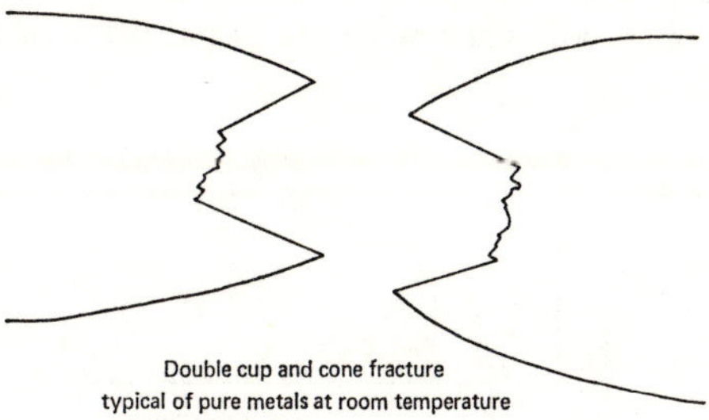

Double cup and cone fracture typical of pure metals at room temperature

(a)

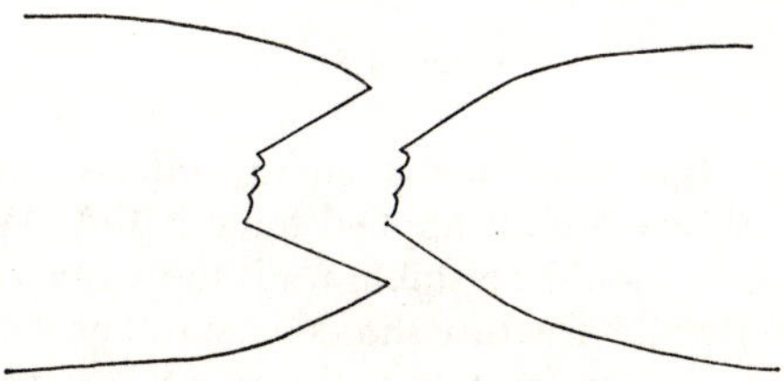

Cup and cone fracture typical of many common alloys

(b)

Figure 4.3

Many single crystals, especially of metals, though fracturing in a brittle manner, do undergo some plastic deformation before fracture occurs. In these cases Sohncke's Law does not, in general, hold.

In polycrystals the fracture surface is macroscopically the plane of maximum tensile stress, though individual grains break on crystallographic planes, giving a fracture surface consisting of many small plane facets.

4.4 The ductile-brittle transition

Brittle fracture does not occur if a tensile specimen can deform plastically by an appreciable amount. The CRSS for plastic deformation varies with temperature more rapidly than the tensile stress for fracture, so most solids show a transition from brittle behaviour to ductile behaviour as the temperature is raised. This is the ductile-brittle transition.

4.5 The theoretical brittle strength

Consider the brittle fracture of a single crystal under tension. Ideally, brittle fracture would occur when the applied stress reaches the value necessary to pull one plane of particles completely away from an adjacent plane. The interparticle forces F between two adjacent planes of particles will vary with the separation of the planes x as

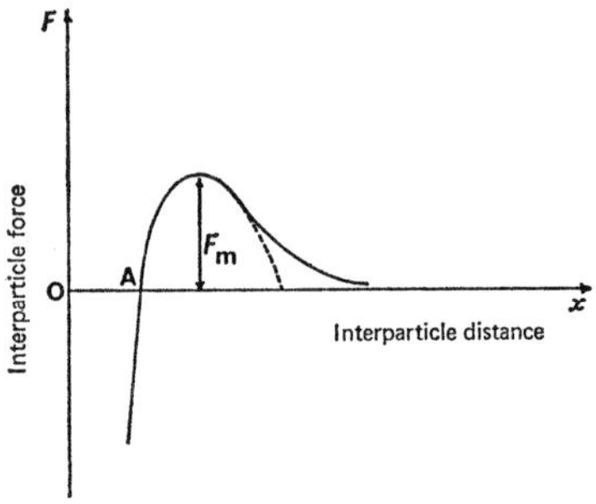

Figure 4.4

shown in fig. 4.4. Ignoring kinetic energy effects, the equilibrium separation in the absence of an applied force is OA. When a force is applied the extension should be stable until the value of F reaches its maximum F_m, after which fracture should occur. The work done by the applied forces in producing fracture is the area under the F–x curve to the right of the point A (repulsive forces are of no interest in this case). Take A as the new origin and assume that the form of the curve is approximately sinsuoidal, as shown by the dotted curve in fig. 4.4.

$$\text{Let } F = F_m \sin kx$$

where k is a constant.

When x is small $F = F_m\, kx$

If the crystal planes being considered are of unit area, F is the applied stress σ, so that

$$\sigma = \sigma_m \sin kx$$

and when x is small $\sigma = \sigma_m\, kx$.

When the separation of two adjacent crystal planes is increased from the equilibrium value a by an amount x, the tensile strain is x/a, and thus, applying Hooke's Law to the case when the strain is small,

$$\sigma = E\frac{x}{a} = \sigma_m\, kx$$

$$\text{so that } k = \frac{E}{\sigma_m a}$$

E is the value of Young's modulus in the direction of tension. Thus,

$$\sigma = \sigma_m \sin\left(\frac{Ex}{\sigma_m a}\right)$$

and hence, the work needed to separate completely the two crystal planes is given by

$$\int \sigma\, dx = \int_0^{\pi} \sigma_m \sin\left(\frac{Ex}{\sigma_m a}\right) dx$$

$$= \sigma_m \left[-\frac{\sigma_m a}{E} \cos\left(\frac{Ex}{\sigma_m a}\right)\right]_0^{\pi} = \frac{2\sigma_m^2 a}{E}$$

The work done is used to create two new surfaces, each of unit area, and each having a free surface energy* γ, so that

$$\frac{2\sigma_m^2 a}{E} = 2\gamma$$

$$\text{or, } \sigma_m = \left(\frac{\gamma E}{a}\right)^{\frac{1}{2}}$$

This approximate calculation gives values for σ_m about one thousand times those obtained experimentally. Griffith assumed that the discrepancy between the theoretical estimates and the observed values of the brittle strength was due to the presence of minute cracks or other flaws, around which a very strong stress concentration resulted when the solid was stressed, and that fracture occurred through the movement of one of these cracks under an applied stress less than the theoretical brittle strength. Most brittle fractures take place rapidly once the crack starts to grow in length, and in these cases the energy needed to propagate the crack is obtained from the elastic strain energy stored in the specimen at the instant that the crack starts to move. Griffith assumed that the crack spreads rapidly through the material if the rate of release of elastic strain energy stored in the material is at least equal to the rate of increase in free surface energy resulting from the production of new areas of crack surface.

Consider the brittle fracture of a large, thin plate of isotropic material containing an infinitely sharp crack of length $2l$, perpendicular to the tensile axis, as in fig. 4.5. Assume that all deformation is elastic, that is, that the fracture is truly brittle. It can be shown that if the crack extends by an amount δl at each tip, under conditions such that the applied forces do no work, the accompanying release of elastic energy is

$$\frac{2\pi\sigma^2 l\delta l}{E}$$

* This concept is discussed in chapter 7.

per unit width of crack. At the same time the new crack surfaces gain energy $4\gamma\delta l$ per unit width. Thus, the total change in energy due to the increase in length of the crack is

$$4\gamma\delta l - \frac{2\pi\sigma^2 l\delta l}{E}$$

and the crack extends if

$$\frac{2\pi\sigma^2 l\delta l}{E} \geqslant 4\gamma\delta l$$

Thus, the critical crack length is

$$l_c = \frac{2\gamma E}{\pi\sigma^2}$$

or, the fracture strength is given by

$$\sigma_c = \left(\frac{2\gamma E}{\pi l_c}\right)^{\frac{1}{2}}$$

When σ_c is equal to the stress needed to make the largest crack grow the fracture occurs.

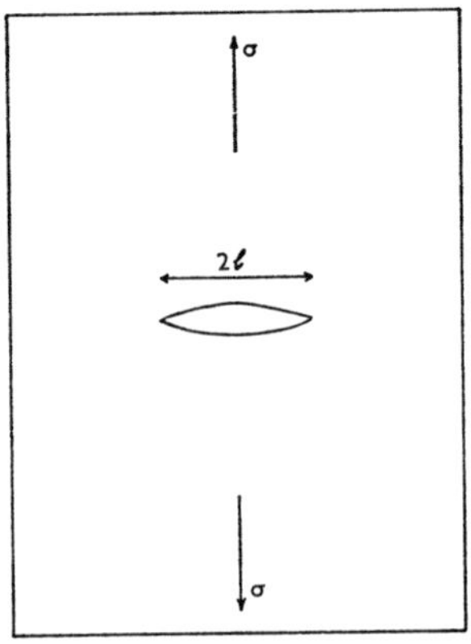

Figure 4.5

It should be possible to increase the brittle strength by removing or making ineffective the most dangerous cracks. (It should be noted that a surface crack of length l is as effective as an internal crack of length $2l$ in acting as a stress concentrator.) Freshly drawn glass rods show a brittle strength close to the theoretical value, though the high strength falls to the usual value in a few hours. The high strength disappears immediately if the rod is touched with any hard body. These experiments show that, with glasses, the dangerous cracks are at the surface and this is also true for most cases of brittle fracture.

In metals showing brittle behaviour, $2l$ would have to be several millimetres. No visible pre-existing cracks or flaws of this size are observed in brittle metals, yet the brittle strength is still low. However, it has been assumed that the deformation during the spreading of the crack is purely elastic, whereas in metals some plastic deformation occurs before the material fractures. To take plastic deformation into account, γ has added to it a term representing the work to produce plastic deformation of the material round the crack, the sum of these being known as the 'surface work parameter' 7.

Then,
$$\sigma_c = \left(\frac{27E}{\pi l_c}\right)^{\frac{1}{2}}$$

The small cracks that spread rapidly under a stress to give brittle fracture are not present in 'as-grown' metals, and must be formed at some stage of the deformation. They may result from stress concentrations set up around inclusions during deformation, or from the coalescence of edge dislocations.

In ductile materials sharp cracks are blunted by plastic deformation, which is localized at the tip of the crack, and this renders the crack ineffective, because the stress concentration depends entirely on the sharpness of the crack tip.

4.6 The plastic deformation of brittle solids

Many normally brittle solids can undergo considerable plastic deformation when, in addition to the axial stress, they are subjected to a hydrostatic pressure. If the brittle strength is σ_m and the specimen has a hydrostatic pressure p applied to it, the tensile stress needed to produce

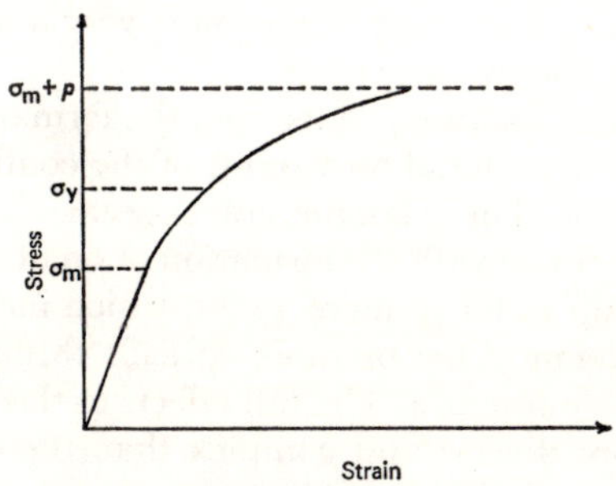

Figure 4.6

brittle fracture is (σ_m+p). This tensile stress gives rise to a shear stress and, if this shear stress reaches the CRSS, corresponding to a tensile stress σ_y, before (σ_m+p) is reached (see fig. 4.6), the material will deform plastically without brittle fracture taking place.

This change in behaviour under a confining pressure accounts for the plastic deformation of many materials normally considered brittle. For example, many rocks deform plastically under a sufficiently high confining pressure. Limestone deforms plastically if the confining pressure exceeds $0{\cdot}6 \times 10^9$ N m^{-2}. This pressure is found at depths of 20 km and over in the earth's crust.

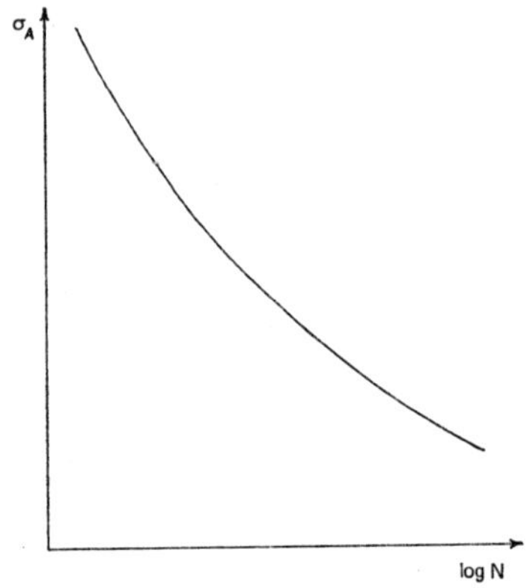

Figure 4.7

4.7 Fatigue fracture

Fatigue is the name given to the inability of a solid to withstand fluctuating stresses. Fracture can be produced by smaller and smaller stresses if these are applied and removed sufficiently often. The maximum stress in a cycle of stressing is less (often very much less) than that needed to produce fracture in tension.

Essentially, fatigue fracture results from the formation and propagation of a crack in the material as a result of the continued application of a fluctuating stress. These fatigue cracks seem to be initiated early in the life of the specimen with the formation of small voids. These voids multiply and link up to form microcracks, which increase in size with continued stress-cycling. One or more spreads through the solid and finally gives complete fracture. The full effects of this mode of stressing a material are more diverse and complex than the outline just given and are, as yet, not well understood.

In most practical cases the usual method of testing is to subject an actual component to the stress cycle it experiences in service, determine the number of cycles of stress needed to cause fracture, and notice where the important fatigue cracks form. If the fatigue life is not acceptable, the component must be redesigned.

However, to attempt to characterize the fatigue resistance of a particular material, it is usual to subject a suitable specimen to a standard test. This is usually an axial test on a cylindrical specimen, in which a given stress cycle is repeatedly applied and the number of cycles N to cause failure is measured. If σ_M is the mean axial stress applied to the specimen and $2\sigma_A$ is the range of stress variation, the applied stress is $\sigma_M \pm \sigma_A$. When σ_M is zero the stress cycle is termed 'alternating' and

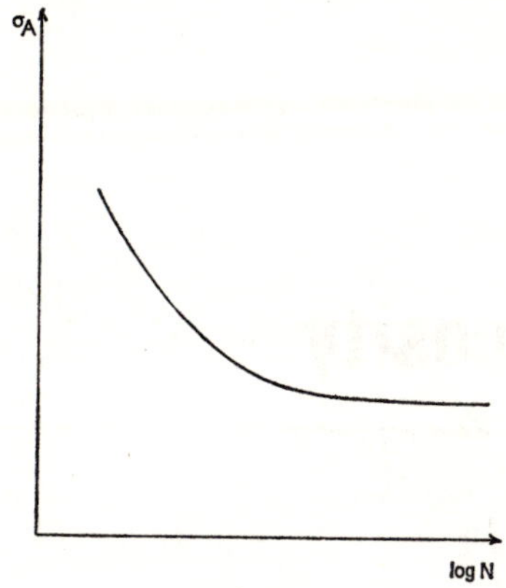

Figure 4.8

when is not zero the term 'fluctuating' is used. The results of tests are usually plotted on a stress-endurance curve (usually known as the *S–N* curve), in which σ_A is plotted against $\log N$. The details of the $\sigma_A - \log N$ curve vary from material to material and also, for a given material, depend on the temperature. For example, many non-ferrous metals at room temperature give a curve of the form shown in fig. 4.7, from which it seems that a finite number of cycles of stressing (albeit a very large number of cycles) will produce fracture, even when the stress amplitude is very small. In contrast, the curve for mild steel, also at room temperature (fig. 4.8), shows a fairly sharp discontinuity, and a definite fatigue limit, that is, a limiting alternating stress amplitude below which the fatigue life seems indefinitely long. This range of stress amplitude is sometimes referred to as the 'pseudo-safe' range of behaviour. Curves of the form of fig. 4.8 are typical of materials that show a sharp yield point in tension, as in fig. 3.5. The curve of fig. 4.7 is typical of materials that have a stress-strain curve in tension of the form of fig. 3.3.

Further reading

Gordon, J. E., *The New Science of Strong Materials*, Penguin Books Ltd, 1968, ch. 3–5.

5
liquid viscosity

5.1 Introduction

A liquid has been defined as an aggregation of basic particles, frequently, in fact, molecules, that possesses a definite volume at a given temperature, but no definite shape; it takes up the shape of the containing vessel. This process may be accomplished very rapidly, as in the case of water, or it may take considerable time, as in the case of thick treacle. When a liquid suffers a continuous deformation under the action of gravity, or of an externally applied force, the process is called 'flow'. As the flow process takes time to be completed, the liquid must be offering some resistance to flow, which may be called quite generally an internal friction or liquid friction.

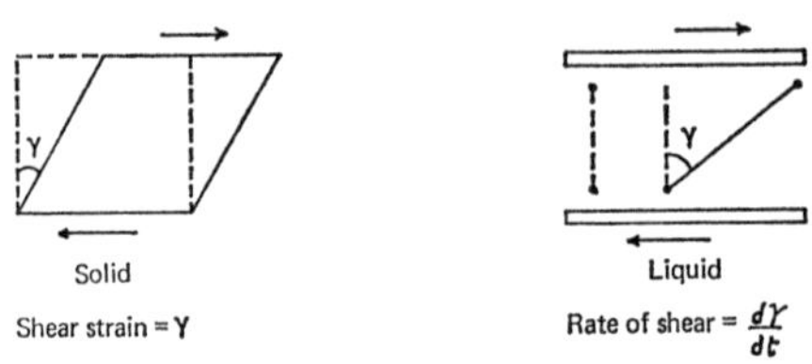

Figure 5.1

Isotherms on the pressure–volume graph for a liquid are practically straight lines, almost parallel to the pressure axis. Thus, liquids are almost incompressible, so that a change of shape implies a shear deformation, and hence a shear stress must be acting. Thus, a liquid deforms continuously under the action of a shearing stress or, in other words, a liquid is unable to sustain a shear stress. This may be contrasted with the behaviour of a solid under the action of a shear stress. When a solid is deformed under the action of a shear stress system, a definite shear strain is produced. If the deformation is purely elastic this shear strain is proportional to the shear stress. On the other hand, when a shear stress is applied to a liquid, the latter deforms continuously at a given rate, and the behaviour of the liquid may be described in terms of the shear stress and the shear strain rate or rate of shear (see fig. 5.1).

When the shear stress is removed from an elastically deformed solid, the solid recovers its original shape completely, but no such recovery occurs in a liquid; the work done in producing a given rate of shear is completely dissipated. The resistance offered by a liquid when it is subject to a continuous shearing is called a viscous force, and the liquid is said to possess viscosity.

Liquid flow under a pressure gradient can occur in two basic ways. The liquid may flow so that there is complete mixing due to the setting up of a very large number of eddy currents in the liquid, or the flow can be essentially in layers, without mixing on the macroscopic scale. Under steady conditions this latter type of flow is quite stable. The flow in which there is complete mixing is referred to as 'turbulent' flow, and for a given liquid is characteristic of high flow rates in broad channels. In a region of turbulent flow all parts of the liquid have the same average flow velocity. Flow in layers is termed 'laminar' or 'streamline' flow, and is typical of low flow rates in narrow channels. The term 'streamline' is also used to describe a line drawn at any instant, such that the tangent at any point is the direction of the velocity vector at that point. When conditions are steady, the velocity of flow at any point is constant and the streamlines do not vary with time. Streamlines can be observed under steady conditions by scattering small, light particles into the liquid and following their paths.

In a liquid in which the flow is laminar, adjacent layers of liquid travel with different speeds. The layer immediately in contact with the solid containing surface is, in fact, adsorbed to that surface, and is therefore, from a macroscopic point of view, at rest relative to the surface. This adsorbed layer is in dynamic equilibrium, as molecules are continually being removed as a result of collisions with other molecules, and different molecules take their place.

5.2 Coefficient of viscosity

Consider a liquid flowing in a streamline manner. Let α and β be, respectively, two streamlines in a very small region of a liquid flowing in layers under steady conditions (see fig. 5.2a). In this region imagine two light particles A and B placed, one on each streamline, at a time t, in such positions that the line joining them is at right angles to the streamlines (which must be closely parallel in this small region of flowing liquid) as in fig. 5.2a. The speed of liquid flow will be slightly different along the two streamlines so that, after a further time δt, the line joining AB makes an angle $\delta\phi$ with its direction at time t. The shear that the liquid between the streamlines has suffered is $\delta\phi$ and $\delta\phi/\delta t$ measures the rate of shear.

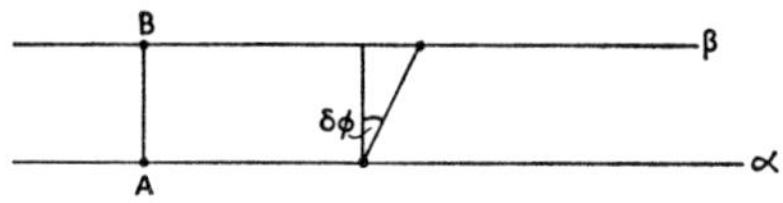

Figure 5.2a

If the streamlines in this small region of liquid are straight and parallel and a distance δx apart, the velocities of flow are constant along each, and the only variation in velocity is in a direction normal to the direction of flow.

Let the velocity of liquid flow along streamline α be u. Then, that along β will be $u+\frac{du}{dx}\,\delta x$, where x is measured normal to the direction of flow. In a small interval of time δt, particle A moves a distance $u\delta t$ and B moves a distance $\left[u+\frac{du}{dx}\,\delta x\right]\delta t$. If the line joining AB was normal to the flow direction at time t, then, after a further time δt, the line makes an angle $\delta\varphi$ with its direction at time t, where

$$\delta\varphi = \left(\frac{du}{dx}\right)\frac{\delta x\,\delta t}{\delta x} = \frac{du}{dx}\,\delta t$$

so, in the limit,

$$\frac{d\varphi}{dt} = \frac{du}{dx}$$

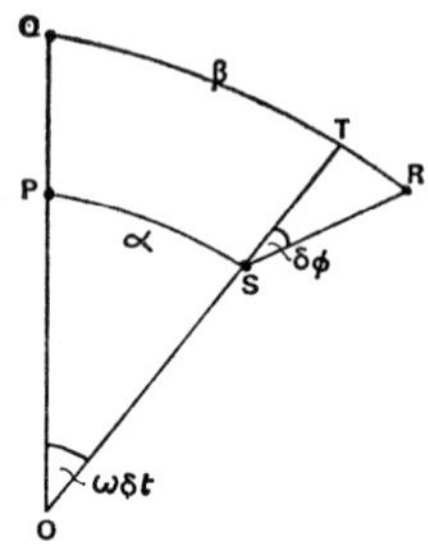

Figure 5.2b

In this case the rate of shear of the liquid is equal to the velocity gradient normal to the direction of flow.

On the other hand, if the streamlines are concentric circles, the situation in the small element of liquid is as shown in fig. 5.2b. At time t

the two light particles are, respectively at P and Q on two streamlines a distance δr apart. O is the centre of curvature of the streamlines and OP $= r$. If the angular velocity of the particle on streamline α is ω that of the particle on streamline β is $\omega + \frac{d\omega}{dr}\,\delta r$.

In a small time δt the particle on streamline α moves from P to S and that on streamline β moves from Q to R where

$$\text{PS} = r\omega\,\delta t, \text{ and QR} = (r+\delta r)\left(\omega + \frac{d\omega}{dr}\,\delta r\right)\delta t.$$

The shear of the liquid that has occurred is $\delta\varphi = \hat{\text{TSR}}$ which is approximately RT/ST.

Thus, $$\delta\varphi = \frac{\text{RT}}{\text{ST}} = \frac{\text{QR}-\text{QT}}{\text{ST}}$$

Now $\text{QR} = r\omega\,\delta t + r\frac{d\omega}{dr}\,\delta r\,\delta t + \omega\delta r\,\delta t$

neglecting $\delta t\,(\delta r)^2$, and

$$\text{QT} = (r+\delta r)\,\omega\,\delta t = r\omega\,\delta t + \omega\,\delta r\,\delta t$$

Hence, $$\delta\varphi = r\,\frac{d\omega}{dr}\,\frac{\delta r\,\delta t}{\delta r} = r\,\frac{d\omega}{dr}\,\delta t$$

and, in the limit, $$\frac{d\varphi}{dt} = r\,\frac{d\omega}{dr}$$

which is the rate of shear in this case.

For many liquids undergoing laminar flow the shear stress acting over surfaces parallel to the direction of flow needed to maintain a given shear rate is proportional to that shear rate. Then

$$\tau \propto \frac{d\varphi}{dt}$$

or, $$\tau = \eta\,\frac{d\varphi}{dt}$$

where η is a constant for a given liquid at a given temperature and pressure. A liquid for which the above relationship holds is termed 'Newtonian', and η is known as the 'coefficient of viscosity' or, more correctly, as the 'coefficient of shear (or dynamic) viscosity'.

In the case where the streamlines are straight, $\eta = \tau \div du/dx$, where τ is the shear stress between two planes parallel with the direction of flow and du/dx is the velocity gradient at right angles to the direction of flow. η has dimensions $\text{ML}^{-1}\,\text{T}^{-1}$. Its SI unit is N s m^{-2}. When a liquid starts to flow under a pressure gradient, the bulk of the liquid begins to move, but the layer adsorbed to the containing wall does not. It tries to prevent the adjacent layers of liquid from moving, and in this process extracts from them momentum which is transferred to the solid wall. If the liquid is pictured as moving in layers, each layer loses momentum to the adjacent layer nearer to the wall and gains momentum from the adjacent layer farther from the wall. Thus, there is a transfer of momentum from layer to layer, and finally the momentum is lost to the wall.

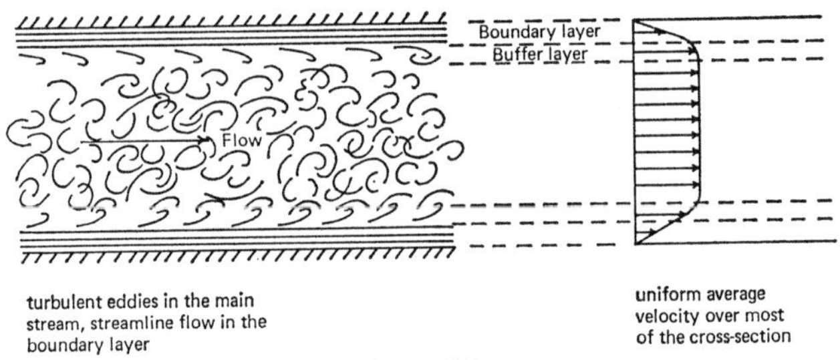

Figure 5.3

Hence, the flow velocity decreases as the wall is approached and tends to zero at the wall. The moving liquid tries to drag the adsorbed layer, and hence the solid wall, along with it, but this tendency is resisted by the rigidity of the wall.

An adsorbed layer exists even when the liquid flow is turbulent. In this case the adsorbed layer maintains a narrow 'boundary layer' in which the liquid flow is laminar even though the bulk flow is turbulent. These two regions are separated by a narrow 'buffer layer' in which the flow is neither truly laminar nor truly turbulent. The liquid velocity increases monotonically with distance from the wall where the flow is

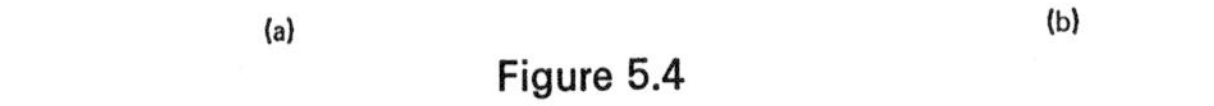

Figure 5.4

streamline, but has a constant mean value where the flow is turbulent. The flow patterns and velocity distributions for liquid flowing in a tube (*a*) under turbulent conditions and (*b*) under streamline conditions are shown schematically in figs. 5.3 and 5.4 respectively.

The quantity η/ρ, where ρ is the liquid density, is called the 'kinematic viscosity', ν, and its SI unit is $m^2\ s^{-1}$.

Examples of Newtonian liquids are water, ether, glycerine, benzene, mercury and most pure liquids. Values for the coefficient of viscosity measured at 15°C are given below.

Liquid	Viscosity (N s m^{-2})
Water	$1{\cdot}1\times10^{-3}$
Ether	$0{\cdot}2\times10^{-3}$
Glycerine	$1{\cdot}5$
Benzene	$0{\cdot}7\times10^{-3}$
Mercury	$1{\cdot}5\times10^{-3}$

5.3 Critical velocity

When a liquid flows in a channel the flow is laminar for very small velocities; it remains laminar as the flow velocity is increased, but, at a certain fairly definite mean flow velocity* in the channel, turbulence sets it. The mean velocity at which this happens is called the 'critical velocity' v_c. When turbulence occurs the definition of coefficient of viscosity is no longer applicable; η is not defined under turbulent conditions.

When the liquid flow is streamline, most of the work done by the applied shear forces is dissipated in overcoming the viscous drag between liquid layers moving at different velocities. However, when the flow is turbulent, most of the energy supplied is used in producing eddies in the liquid. The ease with which eddies are formed is likely to depend on the density of the liquid, since bulk kinetic energy is given to local regions of the liquid. Thus, the critical velocity may be expected to depend on the viscosity of the liquid, its density and on the lateral size, d, of the channel through which the liquid is flowing, since the boundary layer will have an important effect in very narrow channels. Dimensional analysis shows that

$$v_c = \frac{(Re)\,\eta}{\rho\, d} = \frac{(Re)\,\nu}{d}$$

where Re is a dimensionless number known as the Reynolds Number. If d refers to the diameter of circular channels, then turbulent flow occurs for $Re>2000$ and streamline flow for $Re<1000$. When Re lies between 1000 and 2000 the flow is usually streamline if there is no undue disturbance at the entrance to the channel.

If m_c is the critical mass flow rate in a circular channel

$$m_c = \frac{\pi\, d^2\, v_c\, \rho}{4} = \frac{\pi\, d^2\, \rho\, (Re)\,\eta}{4\rho\, d}$$

and hence
$$Re = \frac{4\,m_c}{\pi\, d\,\eta}$$

* Defined as $Q/\pi a^2$ for a cylindrical channel of radius a, where Q is the volume of liquid flowing through the channel per second.

Thus, streamline flow occurs when the mean velocity of bulk flow of the liquid in the channel is less than v_c. The resistance to flow then arises primarily from dragging one liquid layer over the adjacent one, and the shear stress to do this is directly proportional to the rate of shear.

5.4 Liquid models

It was shown in Chapter 1 that in crystalline solids the interparticle forces cause the particles to take up positions in a regular three-dimensional array. Because these particles acquire kinetic energy they oscillate about mean positions and, if the amplitude of oscillation becomes large enough, the particles have sufficient energy to break away from their neighbours, though they then join another particle group. If this happens on a large scale the long-range order typical of a crystal is destroyed and the substance becomes a liquid.

X-ray evidence shows that the molecules in a liquid do have a certain degree of order, and are not arranged entirely at random; any one small cluster of molecules seems to have an almost crystalline order, though the long-range order is absent.

The thermal expansion of most substances on melting is about 5–10%, so that a molecule in a liquid has about as many nearest neighbours as it had in the solid phase. Sublimation corresponds to the breaking of all interparticle bonds in a solid as a particle joins the vapour phase. Since the latent heat of melting is about one-tenth of the latent heat of sublimation, melting corresponds to the breaking of only about one bond with nearest neighbours. Thus, in the liquid the molecules remain in little 'cells' having short-range order, but because of the broken bonds, migration occurs from one cell to an adjacent one. The liquid may then be pictured as a rather broken-up solid. (There are defects, and thus broken bonds, in solids at the melting-point, but these do not destroy the long-range order.) Once a molecule has broken one intermolecular bond by acquiring sufficient kinetic energy, it is able to move to some extent. If it then loses some of its energy, say, in a collision with another molecule, it can form another bond with a different molecule, even though this bond may be short-lived and become broken as the molecule suffers a further collision that increases its energy.

5.5 Liquid flow

The flow of liquids under a pressure gradient can also be described in terms of intermolecular forces or intermolecular bonds. The molecules in the surface of the containing solid are relatively fixed in position, and are able to absorb more kinetic energy in a collision than a molecule in the liquid. They also have broken chemical bonds because they are on the end of the regular three-dimensional array. Thus, when a liquid molecule collides with a wall molecule the production of a chemical bond is more likely than in the case of two liquid molecules colliding. Hence, the solid always has a layer of liquid molecules attached to it—the adsorbed layer. This layer is more or less a monomolecular layer, and is continually exchanging molecules with those in the bulk liquid as a result of collisions.

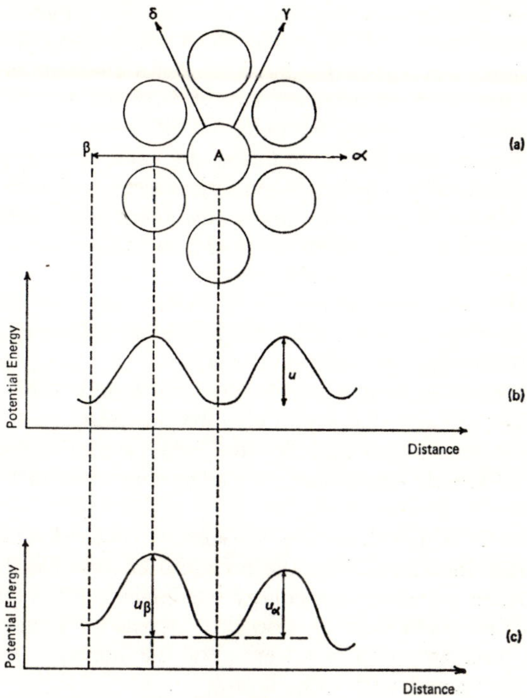

Figure 5.5

If the bulk liquid is caused to flow, the velocity tends to zero at the wall because the adsorbed layer refuses to move. The layer of molecules adjacent to the adsorbed layer is continually exchanging molecules with both the adsorbed layer and the layer on the other side of it as a result of the breaking and formation of chemical bonds. The effect of the applied pressure gradient is to give an overall drift down the pressure gradient to the molecular movements. The resistance to flow observed in liquids, according to this picture, arises from the need to drag one molecule over its neighbour when one (or more) of its chemical bonds has been broken through the molecule obtaining a large amount of kinetic energy.

Consider a very small region of a liquid containing just one cluster of molecules. Over short intervals (that are, nevertheless, long compared with the period of vibration of a molecule) the particular molecules will exhibit short-range order. Fig. 5.5a shows a possible cluster of molecules in two dimensions. The molecules are all in equilibrium positions. For molecule A to move to any of the positions α, β, γ or δ, and so on, it must squeeze past two of its neighbours, and to do this it must acquire sufficient kinetic energy to overcome the intermolecular forces exerted on it by other members of the cluster during such a move. The variation of potential energy of, say, molecule A with respect to movements in the direction $\alpha \rightarrow \beta$ is shown in fig. 5.5b. For a molecule to move from equilibrium site A to an adjacent equilibrium site α it must acquire sufficient energy to overcome the potential hill or barrier between these sites. In a liquid that is not suffering an externally applied stress, molecules occasionally acquire this energy as a result of a kinetic energy fluctuation. Under these conditions, a movement of the molecule from site A to sites β, γ and δ, and so on, is as likely as a move to site α, and thus there is no net flow of liquid. However, if a shear stress is applied to the liquid in the direction $\beta \rightarrow \alpha$, the effect is to tilt the potential energy distribution, as in fig. 5.5c. Then, the jump from site A to site α is more probable than the jump to site β, since the energy barrier u_α is now smaller than u_β and a net liquid flow results. This qualitative treatment shows clearly that liquid viscosity, or resistance to flow, arises from intermolecular forces, and also that, when a shear stress is applied, fluctuations of kinetic energy are able to produce flow in the direction of shear.

As the kinetic energy decreases, the shear stress needed to maintain a given flow rate increases, since the potential energy distribution must be tilted further to reduce the energy barrier in the direction of flow, and hence the coefficient of viscosity increases. Thus viscosity is expected to increase with a lowering of the temperature, and this behaviour is observed with most liquids.

The intermolecular forces are strongly dependent on intermolecular separation. If this distance is changed, there should be a change in the force needed to drag one layer of molecules over another, that is, the coefficient of viscosity should be pressure dependent. Compressing a liquid raises the height of the potential energy barrier, which leads to an increase in viscosity, as is observed.

5.6 The temperature dependence of viscosity

In this section a simple attempt is made to predict the temperature dependence of viscosity on the basis of the model in the previous section.

Referring to fig. 5.5b, let the height of the potential energy barrier be u. The molecule at A tries to climb over this barrier ν_0 times per second, where ν_0 is its vibration frequency. The probability that the molecule A acquires sufficient energy, as a result of a fluctuation, to climb over the potential energy hill can be shown to be

$$\exp(-u/kT),$$

where k is Boltzmann's constant and T is the Kelvin temperature.

Then, the number of successful attempts to climb the potential hill, the 'jump frequency', is

$$\nu = \nu_0 \exp(-u/kT).$$

When a shear stress is applied, the effect is to tilt the potential energy-distance curve, as in fig. 5.5c. This makes jumps from A to α more probable, so that, when the stress is applied, there is a net flow of molecules in the direction of the shear stress. The jump frequency is only slightly changed by the application of moderate stresses, that is, the tilting of the potential energy distribution is quite small, so that the viscosity, which will vary inversely as the jump frequency, will show a temperature variation of the form

$$\eta = \text{constant}\ \exp(u/kT)$$

or,

$$\ln \eta = A + \frac{B}{T}$$

where A and B are constants. Clearly, this result is not restricted to the particular model for a cluster of molecules in a liquid that has been used here. A similar result will be obtained for any process in which the movement of molecule involves climbing over a potential barrier.

The argument may be developed more precisely as follows. When the applied shear stress is τ, the height of the potential barrier changes by $2p\tau a$ per intermolecular spacing, where p is a constant and $2a$ is the intermolecular spacing. Between sites A and α the change in height of the potential energy hill is $-p\tau a$ and between A and β it is $+p\tau a$. The jump frequency in the direction of the applied shear stress becomes

$$\nu_0 \exp(-(u-p\tau a)/kT)$$

and in the opposite direction

$$\nu_0 \exp(-(u+p\tau a)/kT)$$

When a molecule completes a movement over one of the potential barriers it travels a distance $2a$ and makes ν such movements per second, so that the mean velocity of movement relative to the adjacent layer is

$$\begin{aligned} v &= 2\nu a \\ &= 2a\nu_0 \exp(-u/kT)\sinh(p\tau a/kT). \end{aligned}$$

The rate of shear is $v/2a$, and hence the viscosity is given by

$$\begin{aligned} \eta &= \frac{\tau}{v/2a} = \frac{2\tau a}{\nu_0 \exp(-u/kT)\sinh(p\tau a/kT)} \\ &= \frac{2\tau a}{\nu_0} \exp(u/kT)(\sinh(p\tau a/kT))^{-1} \end{aligned}$$

Since the potential distribution is only tilted slightly by the applied shear stress, p is very small and $\frac{p\tau a}{kT} \ll 1$

$$\text{Hence, } \eta = \frac{2\tau a}{\nu_0} \times \frac{kT}{p\tau a} \exp(u/kT)$$

$$= \frac{2kT}{p\nu_0} \exp(u/kT)$$

or, $\eta \propto \exp(u/kT)$, since the exponential term is dominant.

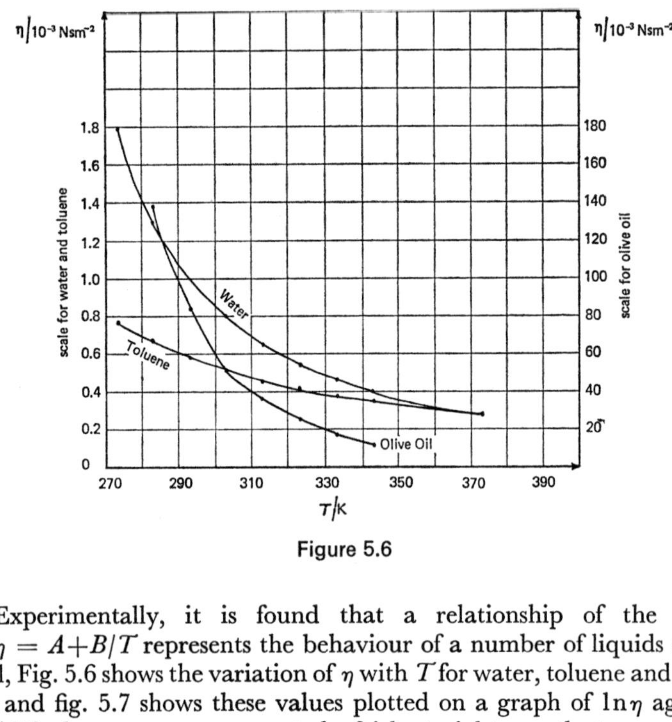

Figure 5.6

Experimentally, it is found that a relationship of the form $\ln \eta = A + B/T$ represents the behaviour of a number of liquids quite well, Fig. 5.6 shows the variation of η with T for water, toluene and olive oil, and fig. 5.7 shows these values plotted on a graph of $\ln \eta$ against $1/T$. The latter curves are seen to be fairly straight over the temperature range considered.

5.7 High rates of shear

To maintain a given macroscopic flow rate, sufficient molecules per second must acquire enough kinetic energy from fluctuations to climb over the potential energy barrier. The maximum rate of flow that can be maintained by the liquid is governed by the jump frequency ν, the number of molecules that cross the potential energy hill per second, and this in its turn cannot exceed ν_0, the vibration frequency of the molecule; except at infinite temperature ν is always much less than ν_0. If the rate of shear made too high, the jump frequency is not high enough to maintain the flow rate, and in this case the substance will

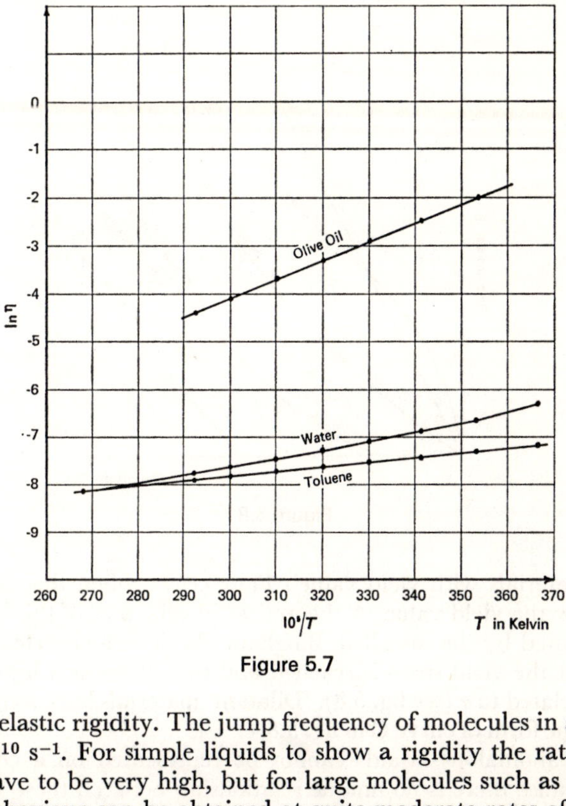

Figure 5.7

show an elastic rigidity. The jump frequency of molecules in a liquid is about 10^{10} s^{-1}. For simple liquids to show a rigidity the rate of shear would have to be very high, but for large molecules such as polymers, elastic behaviour can be obtained at quite moderate rates of shear.

5.8 Non-Newtonian liquids

The liquids that are Newtonian in their behaviour include water, true solutions, simple organic liquids and dilute suspensions. For them a graph of rate of shear $\dot{\epsilon}$ against shear stress τ (known as the 'consistency curve') is a straight line passing through the origin, provided that flow is laminar (see fig. 5.8). Many other liquids, however, do not show this behaviour and are called 'non-Newtonian' liquids. For some of these non-Newtonian or 'anomalous' liquids the behaviour under an applied shear stress can be represented on fig. 5.8 by a curve which may or may not pass through the origin.

A common class of material is the 'pseudo-plastic' material. Its members have a consistency curve of the form of curve A in fig. 5.8. Some pseudo-plastic materials deform like solids until a certain stress Υ, known as the 'yield stress', is applied. For higher shear stresses they behave as normal pseudo-plastic materials and are known as 'pseudo-

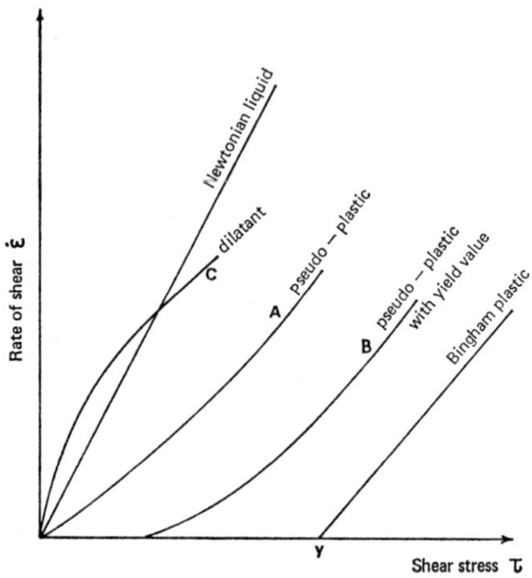

Figure 5.8

plastic materials with yield value'. Flow ceases when the shear stress falls below the yield value. A theoretical idealization of this behaviour is represented by the so-called 'Bingham plastic' which deforms elastically until the yield stress is reached and then flows as a liquid with ϵ linearly related to τ (see fig. 5.8). 'Dilatant' materials have a consistency curve of the form of curve C in fig. 5.8.

Other anomalous liquids cannot be represented on a consistency curve because there is no unique relationship between these variables: their flow characteristics are time dependent. For example, some liquids, when caused to flow at a constant rate of shear, do so under a gradually decreasing shear stress. Paints, inks, glues, plant juices, milk and blood come into this group, the name for which is 'thixotropic'. Other materials exhibit 'inverse thixotropy', and flow easily under low shear rates but become quite rigid at high rates of shear. An example of this group is wet sand, which when walked upon appears to become dry. (Quicksand, on the other hand, is thixotropic.) Another example is a suspension of 43% rice starch in water.

The quantity

$$\frac{\text{shear stress}}{\text{rate of shear}}$$

for any single measurement for an anomalous liquid is termed the 'apparent viscosity'. In general it is not a unique quantity but depends on the rate of shear.

For the Bingham plastic

$$\dot{\epsilon} = \frac{\tau - \Upsilon}{\mu}$$

where μ is known as the Bingham viscosity and Υ is the yield value. Thus,

$$\eta \text{ apparent} = \frac{\tau}{\dot{\epsilon}} = \mu + \frac{\Upsilon}{\dot{\epsilon}}$$

Materials whose behaviour is more or less well represented by this model include slurries, muds, some paints and toothpaste.

Non-Newtonian liquids are characterized by high molecular weight and a readiness to form intermolecular bonds. When a stress is applied is must rise to a value sufficient to assist the molecular kinetic energy to break bonds at a sufficient rate to maintain the imposed rate of flow. This stress may be low or quite large. Molecules of high molecular weight may be of very irregular shape so that several bonds must break before movement can occur, and also molecules may impede one another. After some flow has occurred, the molecules may line up to some degree, so that flow becomes easier and the viscosity decreases.

5.9 Liquid flow through tubes

When a liquid flows through a tube under a pressure gradient, the flow is streamline if the mean flow velocity is less than the critical velocity, and turbulent if it is greater than the critical velocity. When a liquid flows in a streamline manner along a tube of uniform cross-section, and when conditions are steady, the velocity of flow along any streamline remains constant. Any increase in velocity along the tube, together with the very small volume dependence on pressure of a liquid, would mean that the liquid would not fill completely the cross-section of the tube away from the end of entry of the liquid. Such failure to fill the cross-section is not observed in practice. The practical incompressibility of liquids also ensures that no slowing down can occur in the steady state. Thus, when conditions are steady, the force exerted on the liquid by the pressure difference between the ends of the tube is exactly balanced by the viscous forces.

Consider the steady-state streamline flow through a uniform horizontal tube, of inside radius a and length l, of an incompressible Newtonian liquid of density ρ and viscosity η. The velocity of the liquid in contact with the walls (the adsorbed layer) may be taken as zero. Symmetry considerations indicate that the velocity distribution in the liquid will be symmetrical about the axis of the tube, and have its maximum value on this axis. Let the velocity at a distance r from the axis be u.

Consider the forces acting on the liquid in a cylindrical element of radius r, coaxial with the tube. The velocity gradient at the surface of the cylinder is everywhere du/dr, and hence the viscous drag exerted on the surface of this liquid cylinder by the remainder of the liquid is

$$\eta \times 2\pi rl \times \frac{du}{dr}$$

and acts in the opposite direction to the pressure gradient. If, as is assumed here, there is no radial flow of liquid, the pressure will be constant over any particular cross-section. The force acting on the liquid cylinder due to the applied pressure difference is $(p_1-p_2)\pi r^2$, where p_1 is the pressure at the inlet of the tube and p_2 that at the outlet. When conditions are steady these forces must be equal and opposite, that is,

$$(p_1-p_2)\pi r^2 = -\eta 2\pi rl \frac{du}{dr}$$

or,

$$-r\,dr = \frac{2\eta l}{(p_1-p_2)}\,du \tag{5.1}$$

Then, integrating equation (5.1) and using the condition that at the wall of the tube the liquid is at rest, that is, $u = 0$ when $r = a$, gives

$$u = \frac{(p_1-p_2)(a^2-r^2)}{4\eta l} \tag{5.2}$$

This shows that when a liquid flows in a streamline manner along a uniform, horizontal tube the velocity distribution is a paraboloid of revolution with its axis coinciding with that of the tube. When $r = 0$, $u = u_0$, where

$$u_0 = \frac{(p_1-p_2)a^2}{4\eta l}.$$

u_0 cannot be measured directly. Instead, the total volume flow through the tube in a given time must be measured.

The volume flowing per second through an element lying between radii r and $r+dr$ is

$$\begin{aligned} dQ &= 2\pi ru\,dr \\ &= 2\pi(p_1-p_2)r(a^2-r^2)dr/4\eta l \end{aligned}$$

so that the volume flow per second over the total cross-section is

$$Q = \frac{(p_1-p_2)\pi}{2\eta l}\int_0^a r(a^2-r^2)\,dr$$

giving

$$Q = \frac{(p_1-p_2)\pi a^4}{8\eta l} \tag{5.3}$$

Equation (5.3) is known as Poiseuille's equation.

The derivation of Poiseuille's equation assumes that, along the whole length of the tube, the liquid is travelling with constant velocity along any particular streamline. In many experimental arrangements part of the applied pressure difference is used to supply the kinetic energy of the liquid in the tube, and hence a correction must be applied to allow for that part of the pressure difference. This is the 'kinetic energy correction'.

The kinetic energy that the liquid receives per second is

$$\int_0^a \tfrac{1}{2} 2\pi r \rho u \times u^2 dr$$

where p is the density of the liquid. Using equation (5.2) this expression becomes

$$\pi\rho \left(\frac{p_1-p_2}{4\eta l}\right)^3 \int_0^a r(a^2-r^2)^3 dr$$

which equals $$\pi\rho \left(\frac{p_1-p_2}{4\eta l}\right)^3 \frac{a^8}{8}$$

This may be written $$\left(\frac{(p_1-p_2)\pi a^4}{8\eta l}\right)^3 \frac{\rho}{\pi^2 a^4}$$

that is, $$\frac{Q^3 \rho}{\pi^2 a^4}$$

The total work that is done per second by the applied pressure difference is $(p_1-p_2)Q$, so that the work done against viscous forces per second is $(p_1-p_2)Q-\dfrac{Q^3\rho}{\pi^2 a^4}$ and hence the effective pressure difference is

$$(p_1-p_2)-\frac{Q^2\rho}{\pi^2 a^4}$$

Experiment, however, shows this result to be only approximately correct; the second term needs correcting by an arbitrary factor m.

In some experimental arrangements, for example those in which the tube is connected at each end to a reservoir, the above approach is not applicable. However, it is found experimentally that in such cases a correction of the same form is required, so that the effective pressure difference in all cases may be written

$$(p_1-p_2) - \frac{mQ^2\rho}{\pi^2 a^4}$$

where m is a factor whose value is approximately unity, but which must be obtained by calibration in accurate work.

The kinetic energy correction is very important. For example, if water at room temperature is flowing through a tube 1 m long and of radius $0{\cdot}1 \times 10^{-3}$ m under a pressure head of 0·1 m of water, the correction is about 15%. Experiments should be designed and carried out so that the kinetic-energy correction is as small as possible.

At the entrance of the tube the streamlines converge. This involves relative motion of adjacent liquid layers, giving rise to viscous forces that dissipate some of the work that is performed upon the liquid by the pressure difference. Thus, the effective length of the tube is slightly greater than the geometrical length and the correction for this effect, known as the Couette correction, involves replacing l by $(l+na)$. Various values for n have been reported, most of them approximately unity. In most experiments the Couette correction is thus very small.

When both the kinetic energy and Couette corrections are applied to Poiseuille's equation, it becomes

$$\left[(p_1-p_2)-\frac{mQ^2\rho}{\pi^2a^4}\right]\pi a^4 = 8\eta(l+na)Q$$

or,

$$\eta = \frac{\pi a^4(p_1-p_2)}{8Q(l+na)} - \frac{m\rho Q}{8\pi(l+na)} \tag{5.4}$$

m and n can be found experimentally in the following way. Let V be the volume of liquid flowing through the tube in time t, then $Q = V/t$.

Substituting for Q and rearranging, equation (5.4) may be written

$$(p_1-p_2)t = \frac{\eta}{A} + \frac{B}{A} \times \frac{1}{t}$$

where $$A = \frac{\pi a^4}{8V(l+na)} \quad \text{and} \quad B = \frac{m\rho V}{8\pi(l+na)}$$

so that $$\frac{B}{A} = \frac{m\rho V^2}{\pi^2 a^4}$$

For a given apparatus, corresponding values of (p_1-p_2) and t are recorded such that V remains constant. A graph of $(p_1-p_2)\,t$ against $1/t$ gives a straight line of slope B/A and which makes an intercept of η/A on the $(p_1-p_2)t$ axis. m may thus be found from the slope. To determine n the measurements are made with two tubes of different lengths l_1 and l_2. Then, if η/A_1, and η/A_2 are the respective intercepts in these two cases

$$\frac{A_1}{A_2} = \frac{l_2+na}{l_1+na}$$

and hence n may be determined.

An interesting extension of the study of steady liquid flow through a tube under streamline conditions is to the case where the tube is vertical. Consider the arrangement shown in fig. 5.9, in which a uniform vertical tube of length l and internal radius a is joined to a wide vessel in which liquid of density ρ and viscosity η stands to a depth h. The liquid in the tube contained in a central cylindrical column of radius r is acted upon

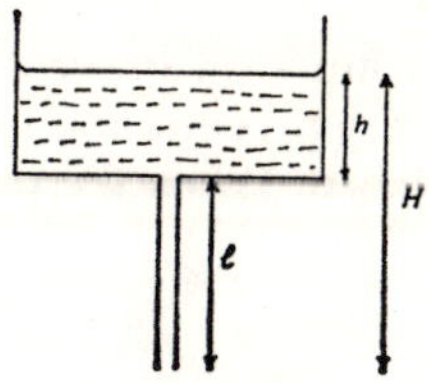

Figure 5.9

by two downwards forces, namely, $\pi r^2 l\rho g$ due to the action of gravity on the liquid column itself, and $\pi r^2(p_1-p_2)$ due to the pressure difference between the ends of the tube, where p_1 is the pressure at the entrance to the tube and p_2 that at the exit. The viscous force, acting over the surface of the cylinder in a direction that opposes the liquid motion, is of magnitude.

$$\eta 2\pi r l \frac{du}{dr}$$

Thus, the equation for steady flow is

$$-\eta 2\pi r l \frac{du}{dr} = \pi r^2 l\rho g + \pi r^2(p_1-p_2)$$

$$-du = r\,\frac{l\rho g+(p_1-p_2)}{2\eta l}$$

Integrating, and remembering that $u = 0$ when $r = a$, gives

$$u = \frac{(a^2-r^2)}{4\eta l}\,(l\rho g+(p_1-p_2))$$

The volume flow of liquid per second through the element lying between r and $r+dr$ is $2\pi r u dr$, so that the total volume flow of liquid per second through the tube is

$$Q = \int_0^a \frac{2\pi r(a^2-r^2)(l\rho g+(p_1-p_2))dr}{4\eta l}$$

$$= \frac{\pi a^4(l\rho g+(p_1-p_2))}{8\eta l} \qquad (5.5)$$

With the arrangement shown $p_1 - p_2$ is, in fact, equal to $h\rho g$ so that

$$Q = \frac{\pi a^4}{8\eta l}(l\rho g + h\rho g) = \frac{\pi a^4}{8\eta l} H\rho g \tag{5.6}$$

5.10 Particles in liquids

If a particle is placed in a liquid of lower density it sinks. At first it accelerates, but the acceleration continuously decreases until a steady or 'terminal' velocity is reached. When the particle is accelerating, the

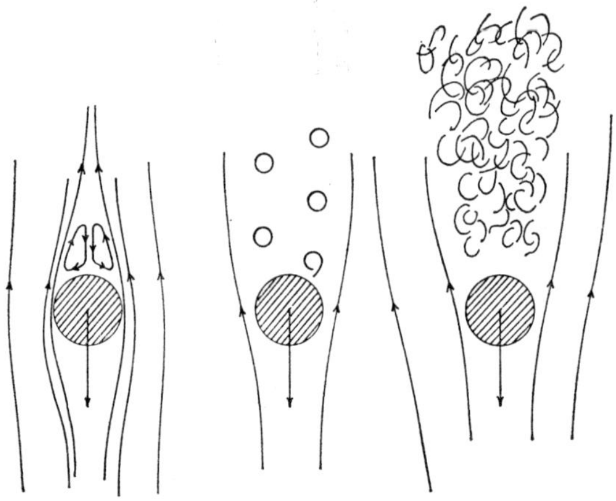

(a) (b) (c)

Figure 5.10

downward force arising from its weight is greater than the sum of the upward forces, namely, the upthrust and the viscous forces arising from the medium. When the terminal velocity is reached the viscous forces and upthrust equal the weight. The nature of the resistive forces will depend on the size, shape, nature of the surface and density of the particle, on the density and viscosity of the liquid and also on the nature of the liquid flow, around the particle as the latter moves. If, in a given liquid, the particle has a smooth shape and surface and is sufficiently light, so that it descends slowly through the liquid, then the resistance to motion arises only from the viscous shear against the particle surface, that is, any fluid flow that occurs is of a streamline nature (shown schematically in fig. 5.10a). The velocity of the liquid flow will be less than the critical velocity.

Even when a body moves slowly through a liquid, so that the movement of the liquid is streamline, there is a region of 'dead' liquid immediately behind the body, in which small eddies are formed (see fig. 5.10a for the case of a sphere). As the velocity increases this eddy formation becomes more pronounced, until a stage is reached when the eddies break away as vortices. Other vortices grow and follow them, forming a 'vortex street'. The vortices are detached alternately from opposite sides of the rear of the body, so that the vortices in the street are staggered (fig. 5.10b). If the velocity of the body increases still more, the isolated vortices disappear to leave an eddying 'wake', which extends for large distances behind the body (fig. 5.10c). Similar behaviour is observed in the case of a body placed rigidly in a steadily flowing liquid.

When the particle is a sphere, the liquid is Newtonian, conditions are such that the particle has reached its terminal velocity, and the flow remains streamline, the situation allows of a simple mathematical treatment. Stokes showed that, under these conditions, the viscous drag exerted by the liquid on the sphere, assuming the liquid to be of infinite extent, is $6\pi\eta r v_t$, where r is the radius of the sphere and v_t is the terminal velocity. If the density of the liquid is ρ and that of the material of the sphere is σ, the Archimedean upthrust on the sphere is $\frac{4}{3}\pi r\rho g$, and the weight of the sphere is $\frac{4}{3}\pi r^3\sigma g$, where g is the intensity of gravity. When the terminal velocity has been reached, the weight of the sphere is equal and opposite to the viscous and upthrust forces, and so

$$6\pi\eta r v_t + \frac{4}{3}\pi r^3\rho g = \frac{4}{3}\pi r^3\sigma g$$

or,

$$\eta = 2r^2 g(\sigma - \rho)/9v_t \tag{5.7}$$

This equation holds for streamline flow in an infinite medium. In practice no medium is of infinite extent, and corrections must be applied. Most measurements and studies are made on spheres falling through liquids contained in cylinders, and the correction is then considered in two parts, namely, the 'end effect', which takes account of the finite length of the cyinder, and the 'wall effect', which allows for the finite radius of the cylinder.

The end effects are not very important if measurements are made only over the middle third of the depth of the liquid. The falling sphere must, of course, have acquired its terminal velocity before it reaches this part of the liquid.

To allow for the wall effect, Ladenburg suggested that the true viscosity was related to that obtained by direct application of equation (5.7) by

$$\eta \text{ true} = \frac{\eta \text{ measured}}{\left(1 + 2{\cdot}4\,\dfrac{r}{R}\right)}$$

where R is the radius of the cylinder containing the liquid. This implies that the terminal velocity is a linear function of r/R. This assumption can be tested by repeating the measurement of v_t in a series of cylinders of different radii. By plotting a graph of v_t measured against r/R, the true value of v_t, that is, the terminal velocity in an infinite liquid, can be obtained from the intercept of the curve with the v_t axis where r/R is zero, as is shown in fig. 5.11.

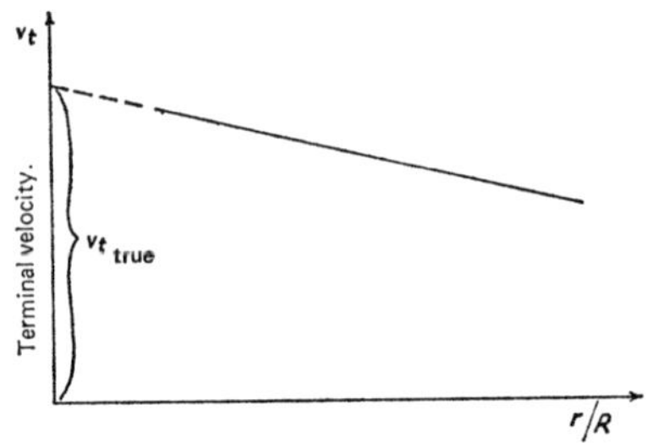

Figure 5.11

5.11 Commercial viscometers

Instruments for measuring viscosities may be relative or absolute, that is, they may need calibrating with liquids of known viscosity, or they may be able to give a value without such calibration. In practice relative instruments are often more convenient to use.

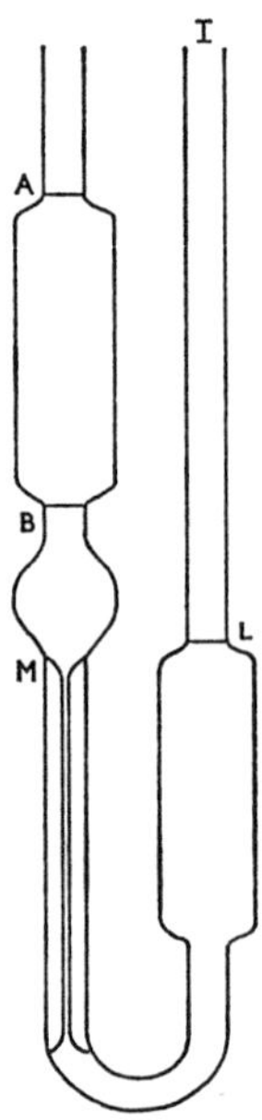

Figure 5.12

A widely used example of the first group is the Ostwald viscometer. This is a U-tube type of instrument in which the pressure gradient is produced by gravity, and one form is shown in fig. 5.12. A specified quantity of liquid is introduced at I, the steady liquid levels being at L and M. The liquid is then drawn into the other limb, so that the upper surface of the liquid is above A, and the time for the liquid level to fall freely between A and B is noted. The arrangement is similar to the vertical tube discussed in section 5.9, but the pressure head decreases with time. The time taken for a given volume of liquid to flow through the capillary section of the tube will increase as the shear viscosity increases, but decrease as the liquid density increases, so that the time t taken for the liquid level to fall from A to B will depend on η/ρ, that is, upon the kinematic viscosity. The instrument may be calibrated by observing t for a number of liquids of known kinematic viscosity ν and plotting a graph of t against ν.

Another example of this class of instrument is the Redwood viscometer, which consists of an open cup, through the centre of the base of which is an accurately bored hole fitted with a valve. The cup is filled to a given mark with liquid, the valve is opened and the time taken for a certain volume of liquid to discharge is noted. This instrument is useful for very viscous liquids; the Ostwald viscometer is used for mobile liquids.

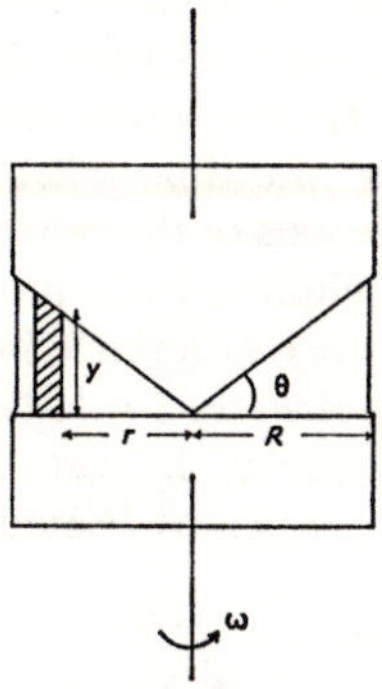

Figure 5.13

In both these viscometers, as in a viscometer based on a sphere falling through a liquid, different parts of the liquid undergo different rates of shear. This does not matter in the case of Newtonian liquids, but for non-Newtonian liquids the apparent viscosity depends on the rate of shear. In such cases it is more meaningful to use a viscometer in which all parts of the liquid suffer the same rate of shear. Such an instrument is the cone and plate viscometer shown in fig. 5.13. A large angle fixed cone has its axis coincident with the axis of rotation of a circular plane surface of radius R and the sample is contained in the space between them.

Consider the liquid contained in the element with its axis parallel to the axis of rotation and its surfaces lying at distances r and $r + dr$ from this axis. If the relative angular velocity of the surface is ω rad s^{-1} the rate of shear of the element is $\omega r/y$. Now y/r equals $\tan\theta$, so that the rate of shear at a distance r from the axis is $\omega/\tan\theta$ which is independent of r. The shearing stress is thus independent of r. When θ is small $y/r \simeq \theta$ and the rate of shear is ω/θ. Let the shear stress be τ. The torque exerted through the element of liquid lying between r and $r + dr$ is then $\tau r 2\pi r dr$ and the total torque T is

$$T = \int_0^R \tau 2\pi r^2 dr = \frac{2\pi\tau R^3}{3}$$

Thus τ is $3T/2\pi R^3$,

and hence the quantity $\dfrac{\text{shear stress}}{\text{rate of shear}}$ is $3T\theta/2\pi R^3\omega$ when θ is small.

For a Newtonian liquid this is the shear viscosity and for a non-Newtonian liquid it is the apparent viscosity.

5.12 Examples

(*a*) A right cylindrical vessel of area of cross-section $0{\cdot}01\,m^2$ has projecting from it a uniform horizontal tube of length $0{\cdot}5$ m and internal diameter $0{\cdot}2 \times 10^{-3}$ m. If the vessel contains water of viscosity $1{\cdot}0\times10^{-3}$N s m^{-2} calculate the time taken for the liquid level to fall from 1 m to $0{\cdot}5$ m above the axis of the horizontal tube.

At any instant t let the height of the liquid level above the axis of the tube by y m. Then the pressure difference between the ends of the tube is $\rho g y$ where ρ is the density of water ($=10^3\,kg\,m^{-3}$). The volume rate of flow of water through the tube at the instant t is (assuming streamline flow and no end corrections) given by Poiseuille's equation

$$Q = \frac{\rho g y \pi a^4}{8\eta l}$$

where the horizontal tube is of length l and internal radius a and the liquid viscosity is η.

Now $Q = dV/dt$ where dV is the volume of liquid flowing through the tube in time dt and $\frac{dV}{dt} = -A\frac{dy}{dt}$ where A is the area of cross-section of the cylinder. The minus sign is introduced since Q is positive, but $\frac{dy}{dt}$ is negative if y is measured upwards from the axis of the tube.

Then,
$$-A\frac{dy}{dt} = \frac{g y \pi a^4 \rho}{8\eta l}$$

or,
$$-\frac{8A\eta l}{\rho g \pi a^4}\frac{dy}{y} = dt$$

If the water level changes from y_1 to y_2 in a time τ

$$\frac{8A\eta l}{\rho g \pi a^4}\ln\left(\frac{y_1}{y_2}\right) = \tau$$

Substituting the values given

$$\tau = \frac{8\times0{\cdot}01\times10^{-3}\times0{\cdot}5\times\ln 2}{10^3\times9{\cdot}81\times\pi\times(0{\cdot}1\times10^{-3})^4}$$
$$= 9{\cdot}0\times10^6\ \text{s}$$

(*b*) A large cylindrical beaker contains water to a depth of $0{\cdot}1$ m. Spherical particles of emery of density $4{\cdot}0\times10^3$ kg m^{-3} of various sizes are stirred up in the water and allowed to settle. Calculate the size of the largest particle still in suspension 1 hour after stirring ceases. The viscosity of water at the temperature of the experiment is

$$1{\cdot}0\times10^{-3}\ \mathrm{N\ s\ m^{-2}}$$

Assume that end effects and wall effects are negligible. The largest particle still in suspension at the end of 1 hour will have travelled from the surface through 0·1 m of water in that time. Assume that the terminal velocity is reached almost immediately the particle begins to descend. Then the terminal velocity is

$$\frac{0{\cdot}1}{60\times60}\ \mathrm{m\ s^{-1}}$$

Substituting in equation 5.7 gives

$$10^{-3} = \frac{2\times r^2\times9{\cdot}81\times10^{-3}(4-1)\times60\times60}{9\times0{\cdot}1}$$

and thus $r = 2{\cdot}1\times10^{-6}$ m

Problems

Assume all liquids are Newtonian.

1. Find the shear stress needed to slide a flat plate at the rate of $0{\cdot}07\mathrm{m\ s^{-1}}$ over a large flat surface separated from it by a film of oil of uniform thickness $0{\cdot}3\times10^{-3}$m and having a viscosity $0{\cdot}45\ \mathrm{N\ s\ m^{-2}}$.

2. Water flows through a horizontal uniform tube 0·4 m long and of internal radius $0{\cdot}50\times10^{-3}$ m at a rate of $7{\cdot}6\times10^{-6}\ \mathrm{m^3\ min^{-1}}$ when the pressure head is 0·3 m of water. Show that the liquid flow is streamline and calculate the viscosity of water at the temperature of the experiment (*a*) without applying any corrections to Poiseuille's equation, (*b*) applying the kinetic energy correction with $m = 1$.

3. Two vessels of equal area of cross-section A containing liquid of density ρ are joined near their bases by a uniform horizontal capillary tube of length l and internal radius r. Initially the liquid surfaces are at heights $3h$ and h, respectively, above the axis of the horizontal tube. Calculate the time taken for the difference in level to become h if the coefficient of viscosity of the liquid is η and the liquid flow is streamline.

4. A spherical ball bearing of radius $0{\cdot}5\times10^{-3}$ m made of material of density $7{\cdot}9\times10^{-3}\ \mathrm{kg\ m^{-3}}$ falls along the axis of a vertical right circular cylinder containing liquid of density $1{\cdot}26\times10^{3}\ \mathrm{kg\ m^{-3}}$ and viscosity η. When the ball bearing has reached its terminal velocity it takes 68·3 seconds to travel a 0·15 m length of the middle section of a cylinder of diameter 0·02 m and 64·6 seconds to travel the same distance when the cylinder diameter is 0·04 m. Calculate η.

5. Two equal circular discs A and B of radius r are placed separately in a liquid of viscosity η. The discs are in parallel planes at a small

separation d and are so positioned that a line drawn through the centre of A and perpendicular to its plane passes through the centre of B. Disc A is caused to rotate about this line as axis with constant angular velocity ω. Find the couple needed to keep B at rest, neglecting any edge effects.

6. A hollow right circular cylinder has an internal radius b. A second right circular cylinder of external radius $a (b > a)$ is made coaxial with respect to the first. Each cylinder is of length l and the space between them is filled with a liquid of viscosity η. If the outer cylinder is fixed show that, neglecting end corrections, the steady couple T needed to rotate the inner cylinder at a constant angular velocity ω_0 is given by

$$4\pi l \omega_0 \eta = T\left(\frac{1}{a^2} - \frac{1}{b^2}\right).$$

(This is the basis of the Searle viscometer.)

Further reading

Dinsdale, A., and Moore, F., *Viscosity and its Measurement*, The Institute of Physics and the Physical Society Monographs for Students, 1962.
Pryde, J. A., *The Liquid State*, Hutchinson, London, 1966.

6
the viscosity of gases

6.1 Introduction

It has been shown in section 1.2 that intermolecular forces are short range. When a substance is in the gaseous phase the molecules are, for most of the time, so far apart that the forces between them are negligible. However, when the gas molecules collide, either with other gas molecules or with the walls of the container, large repulsive forces are brought into play. Because of the very short-range nature of the intermolecular forces, the mean potential energy of the interaction between the molecules of a gas is almost zero, except when the gas is considerably compressed. Another consequence is that the 'gas laws' do not depend on the nature of the material of which the walls of the container are made.

At each collision the direction of motion of a molecule is abruptly changed, so that it suffers an abrupt change of momentum. This may be accompanied by a change in kinetic energy, though, if the temperature of the gas remains constant, the total kinetic energy of all the molecules remains constant. Thus, the effects of inelastic and superelastic collisions balance so that, on average, collisions may be treated as elastic. Because of the very small size of molecules, even in a gas at low pressure, a very large number of molecules strikes unit area of the container wall per second. The pressure exerted by the gas equals the rate of change of the momentum of the molecules striking unit area of the wall. The wall suffers a momentum change per unit area per second equal in magnitude to that experienced by the gas striking that area per second, and may be pictured as first exerting a force sufficient to reduce to zero the velocities of the incident gas molecules, and then giving them a velocity in a direction away from the wall.

Experiment shows that gas molecules are scattered in a random manner after collision with the wall of the container. (Since the incident directions are random, this is equivalent, from the standpoint of momentum changes, to treating all the molecules as if they suffered regular reflection from the wall.) This result is consistent with Langmuir's view that molecules striking the container wall are, in fact, adsorbed and re-emitted to the bulk of the gas at a later time. The direction of motion on re-emission would bear no relation to the incident direction.

On the basis of the molecular picture of a gas, and after making certain assumptions, an expression for the pressure p exerted by the gas can be derived. The necessary assumptions are (*a*) the molecules are randomly scattered as a result of collisions with the walls of the container; (*b*) collisions between molecules and between molecules and the wall must, on average, be elastic; (*c*) the molecules only exert forces on each other and on the walls during collisions (this is equivalent to taking the form of the intermolecular force–intermolecular distance curve to be that of fig. 6.2;) (*d*) the volume occupied by the molecules is negligible, that is, x_o in fig. 6.2 is made vanishingly small.

Let the gas consist of identical molecules, each of mass m, and let there be n molecules per unit volume. These molecules can be divided into groups according to the magnitude of their velocities. Let group 1 consist of n_1 molecules, each having a velocity c_1, with components u_1, v_1, w_1 in the x, y, z directions respectively. Group 2 will consist of n_2 molecules, each with velocity c_2, having components u_2, v_2, w_2 and so on. Consider the molecules striking a small element of area dS perpendicular to the x-direction in a time dt. Since the distribution of molecular velocities as a whole is random, at any instant half the molecules with velocity components u_1, v_1, w_1, will be moving towards dS, and half will be moving away. The number of molecules of group 1 striking the area dS in time dt will be one-half those contained in a cylindrical volume having a base of area dS and a vertical height $u_1\,dt$, that is, $\frac{1}{2}\,n_1\,u_1\,dS\,dt$. Each molecule has a momentum in a direction perpendicular to dS of mu_1. Thus, in a time dt, the momentum perpendicular to dS brought up to dS by the molecules in group 1 is $\frac{1}{2}\,m\,n_1\,u_1^2\,dS\,dt$. The total momentum in the direction perpendicular to dS brought up to the element in a time dt is obtained by summing over all the groups of molecules and is

$$\tfrac{1}{2}\,m\sum n_1\,u_1^2\,dS\,dt$$

Put

$$\overline{u^2} = \frac{\sum n_1 u_1^2}{\sum n_1} = \frac{n_1u_1^2+n_2u_2^2+n_3u_3^2+\ldots}{n_1+n_2+n_3+\ldots}$$

$$= \frac{n_1u_1^2+n_2u_2^2+n_3u_3^2+\ldots}{n}$$

Then, the total momentum perpendicular to dS brought up to it in time dt by all the molecules is $\frac{1}{2}\,m\,n\,\overline{u^2}\,dS\,dt$.

When conditions are steady there is no net flow of gas in any direction and hence the total momentum in the x-direction must be exactly reversed by collisons with dS. The total change in momentum in the x-direction that occurs in a time dt is thus $m\,n\,\overline{u^2}\,dS\,dt$. This equals the impulse of the force exerted by the molecules on the element of area dS in a time dt. This impulse is also equal to $p\,dS\,dt$ so that

$$p\,dS\,dt = m\,n\,\overline{u^2}\,dS\,dt$$

or,

$$p = m\,n\,\overline{u^2}.$$

Let c_m be the root mean square velocity of the molecules, defined by

$$c_m{}^2 = \frac{n_1c_1{}^2+n_2c_2{}^2+n_3c_3{}^2+\ldots}{n_1+n_2+n_2+\ldots}$$

Then,

$$c_m{}^2 = \frac{n_1u_1{}^2+n_1v_1{}^2+n_1w_1{}^2+n_2u_2{}^2+n_2u_2{}^2+n_2w_2{}^2+n_3u_3{}^2+\ldots}{n}$$

$$= \frac{n_1u_1{}^2+n_2u_2{}^2+\ldots}{n}+\frac{n_1v_1{}^2+n_2v_2{}^2+\ldots}{n}+\frac{n_1w_1{}^2+n_2w_2{}^2+\ldots}{n}$$

$$= \overline{u^2} + \overline{v^2} + \overline{w^2}.$$

Since the motion of the molecules is completely random

$$\overline{u^2} = \overline{v^2} = \overline{w^2}$$

Hence,

$$\overline{u^2} = \tfrac{1}{3}c_m{}^2$$

and

$$p = \tfrac{1}{3}mn\,c^2{}_m = \tfrac{1}{3}\rho\,c_m{}^2 \tag{6.1}$$

where ρ is the density of the gas at the temperature in question.

It has been shown theoretically, and verified experimentally, that the distribution of molecular velocities, at a given temperature, follows closely the distribution shown in fig. 6.1. As the temperature is raised the distribution becomes more scattered and the maximum moves in the direction of higher velocities. The analytical expression for the curve is known as the Maxwell distribution law and has the form

$$\mathcal{N} = 4\pi n\left(\frac{m}{2\pi kT}\right)^{3/2}\exp(-mc^2/2kT)\,c^2dc \tag{6.2}$$

where $\mathcal{N}$ is the number of molecules of mass m per unit volume whose velocities have magnitudes in the range c to $c+dc$, n is the total number of molecules per unit volume, k is Boltzmann's constant and T is the Kelvin temperature.

One important consequence of equation (6.1) is that, for common gases at room temperature, the root mean square velocity, which for the Maxwell distribution is close to the mean velocity,* is very high, about 10^3m s^{-1}.

* It can be shown that the mean velocity is, in fact, $\sqrt{\dfrac{8}{3\pi}}$ times the root mean square velocity.

For real gases the variation of intermolecular force with molecule separation is not as simple as assumed above. The attractive force is not zero, for when the molecules of the gas are brought sufficiently close together (by increasing the pressure) the gas condenses, provided the kinetic energy is not too great, that is, provided the temperature is less

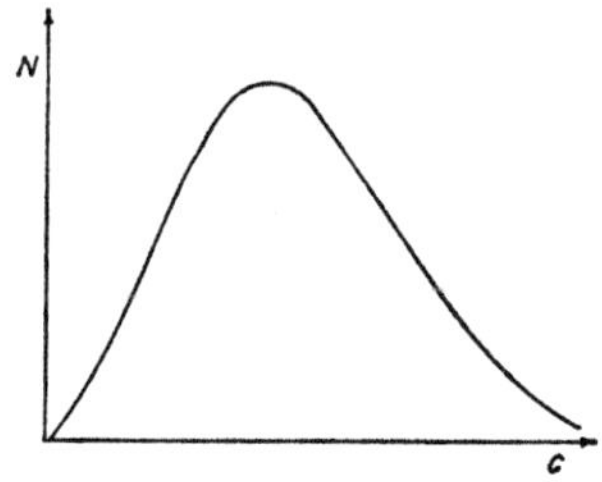

Figure 6.1

than a certain value—the critical temperature. At temperatures above the critical temperature the molecules have sufficient kinetic energy to break away from their neighbours no matter how great the pressure. Also, the repulsive force is such that the centres of the molecules cannot approach closer than a certain finite separation, that is, the molecules do have a finite size.

6.2 The mean free path

The distance that a gas molecule travels, on average, between successive collisions is called the 'mean free path'. Consider a gas in which the molecules are treated as hard spheres of diameter σ (the intermolecular force–intermolecular distance curve will then have the form of fig. 6.2 with x_0 equal to σ) and in which there are n molecules per unit volume. Let a given molecule be moving with a velocity c and imagine that all the other molecules are at rest. If the centre of the moving molecule comes within a distance σ of the centre of one of the stationary molecules, a collision is considered to occur. In unit time the moving molecule travels a distance c, and in so doing collides with all molecules whose centres lie in a volume $\pi\sigma^2 c$. On average there will be $\pi\sigma^2 cn$ molecules with centres in this volume, so that this is the average number of collisions, giving a mean free path

$$L = \frac{c}{\pi\sigma^2 cn} = \frac{1}{\pi\sigma^2 n} \tag{6.3}$$

σ is known as the 'gas kinetic diameter'. Despite the crudeness of this model the result it gives is not altered very much by introducing a number of refinements. For example, taking into account the facts that all the gas molecules are moving, and that they do not all have the same velocity but velocities are distributed according to the Maxwell law, gives

$$L = \frac{1}{\sqrt{2}} \times \frac{1}{\pi\sigma^2 n} \tag{6.4}$$

It will be shown in section 6.3 that values for L can be obtained from measurements of the gas viscosity. At s.t.p. the mean free path for most common gases is of the order of 10^{-7} m.

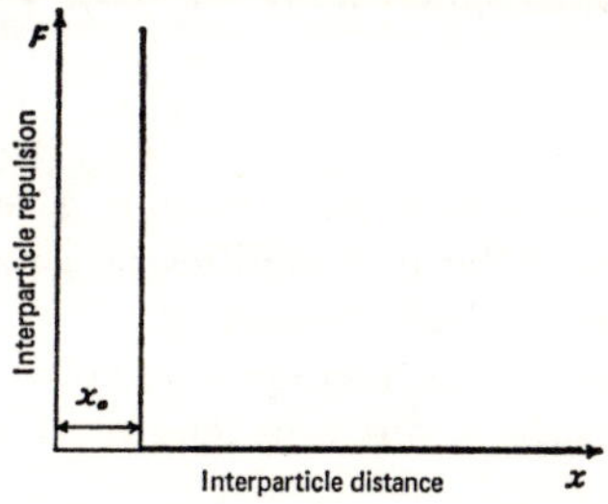

Figure 6.2

Using the values of L, values of σ can be calculated if n is known. The values of σ turn out to be of the order of 10^{-10} m. For example, for hydrogen σ is $0{\cdot}27 \times 10^{-9}$ m.

In practice the molecules are not hard spheres. The intermolecular force curve has the form of fig. 1.2a, and hence two molecules approach one another until their kinetic energy is completely transformed into potential energy of repulsion. Since the kinetic energy of the molecules increases with temperature, a closer approach of the molecules is possible under these conditions, that is, the molecular diameter decreases with increase in temperature.

The free paths in a gas are distributed about the mean free path and the form of this distribution may be found in the following way.

The probability that a collision between two molecules will occur when a molecule travels a very small distance dx is proportional to dx, and may be written $\alpha\ dx$, where α is a constant. Let the probability that a molecule will describe a path of at least a (finite) length x be $f(x)$. The probability that there will not be a collision in a distance dx is $(1-\alpha dx)$, so that the probability that a molecule will travel a distance x and then a distance dx without making a collision is

$$f(x)(1-\alpha dx).$$

From the definition of $f(x)$ this must be equal to $f(x+dx)$, that is, to

$$f(x)+\frac{df(x)}{dx}\ dx$$

Hence, $$f(x)+\frac{df(x)}{dx}\,dx=f(x)-\alpha f(x)\,dx$$

or, $$\frac{df(x)}{f(x)} = -\alpha\,dx$$

Integrating gives $$\ln f(x) = -\alpha x+\text{constant}$$

or, $$f(x) = Ae^{-\alpha x}$$

where A is an arbitrary constant. If x is zero, $f(x)$ must be unity and hence

$$f(x) = e^{-\alpha x}$$

The probability of a free path having a length between x and $x+dx$ is the product of the probabilities of a path of length x and a collision in dx, that is, $f(x)\alpha dx$.

The mean free path is the average value of x weighted according to the number of free paths of different lengths. Thus

$$\text{mean free path } L = \frac{\int_0^\infty x f(x)\alpha\,dx}{\int_0^\infty \alpha f(x)\,dx}$$

$$= \frac{\int_0^\infty x\,e^{-\alpha x}\alpha\,dx}{\int_0^\infty \alpha\,e^{-\alpha x}\,dx} = \frac{1/\alpha^2}{1/\alpha} = \frac{1}{\alpha}$$

Thus, the constant in the expression for the probability is the reciprocal of the mean free path L, so that

$$f(x) = e^{-x/L}.$$

This expression shows that free paths which are much greater than the mean are very infrequent.

6.3 Gas viscosity

The coefficient of viscosity has been defined in terms of the force needed to maintain the steady flow of one layer of fluid over an adjacent layer under conditions of streamline flow. In a gas at rest on the macroscopic scale, the molecules are moving in random directions with high average velocities. One result of causing a bulk flow of gas is to superimpose a very much smaller drift velocity on the random velocity. It is this drift or flow velocity to which reference is made in the definition of viscosity. The random velocity of the molecules means that there is a continuous interchange of molecules between adjacent gas layers, so that some molecules move into regions where the flow velocity is lower. Here they suffer collisions with the more slowly drifting molecules and give momentum to

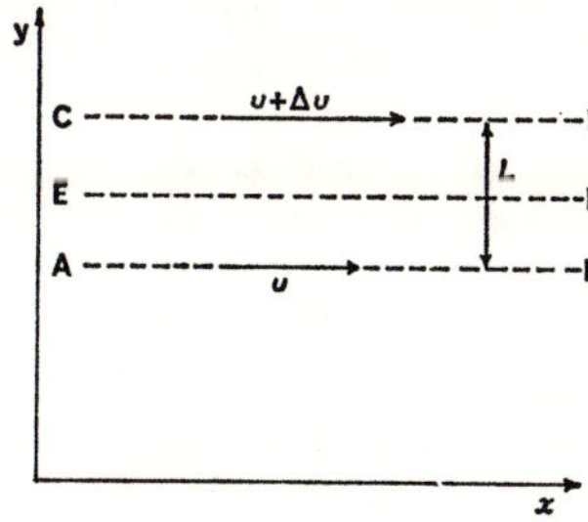

Figure 6.3

these molecules, thus trying to speed them up. Similarly, other molecules move into regions of higher drift velocity and as a result of collisions try to slow down that layer. In this way the exchange of molecules between layers with different drift velocities tries to reduce the relative velocities to zero, so that an applied force is needed to maintain a gas flow.

Let a gas be flowing under streamline conditions in the positive x direction, and let AB (fig. 6.3) represent the plane where the drift velocity is u. Consider the molecules crossing unit area of a reference plane EF. If the number of molecules per unit volume is n and their mean velocity of random motion is $\bar{c}$, since they are moving in random directions the number of molecules crossing unit area of EF in unit time travelling in the positive y direction is $1/6\ n\ \bar{c}$, and the same flux of molecules per unit area is travelling in the negative y direction. The molecules can be considered to have acquired their flow velocity in their last collision, and will change their momentum to that appropriate to the plane in which they find themselves at their next collision. The molecules travelling in the negative y direction made their last collision at a distance L, the mean free path, from AB. After that collision the flow velocity of the molecules is $u+\Delta u$. The momentum carried by molecules moving from the plane AB upwards per unit area per second is $1/6\ nm\bar{c}u$. After collision in the plane CD, the momentum adjusts to $1/6\ nm\bar{c}\ (u+\Delta\ \text{u})$, corresponding to a momentum change of $1/6\ nm\bar{c}\Delta\ u$. Similarly, molecules moving from the plane CD to AB suffer a momentum change of the same magnitude but opposite sign, giving a total momentum change per unit area per second of $\frac{1}{3}mn\bar{c}\ \Delta u$.

Since $$\Delta u = L\frac{du}{dy}$$

where du/dy is the flow velocity gradient in the direction of y, the transfer of momentum per unit area per second is $\frac{m}{3} n \bar{c} L \frac{du}{dy}$ This is equal to τ, the shear stress needed to maintain the relative motion of the layers. The coefficient of viscosity is defined by

$$\tau = \eta \frac{du}{dy}$$

so that

$$\eta \frac{du}{dy} = \frac{mn\bar{c}L}{3} \frac{du}{dy}$$

or,

$$\eta = \frac{mn\bar{c}L}{3} = \frac{\rho\bar{c}L}{3} \qquad (6.5)$$

It may be noticed that, at a given temperature, $\bar{c}$ is a constant, so that $\eta \propto \rho L$.

Now $L \propto 1/\rho$, so that $\eta =$ constant and hence, at a given temperature, η should be independent of pressure. This result is known as Maxwell's law of gaseous viscosity and has been verified experimentally. At very low pressures, however (about 10^{-1} N m^{-2} with 'normal' laboratory apparatus), this law breaks down and viscosity decreases with decreasing pressure. This behaviour occurs when the mean free path becomes so long that it is comparable with the dimensions of the apparatus. Under these circumstances collisions with the container walls are more important than collisions between gas molecules.

6.4 Streamline flow of gas through a cylindrical tube

An expression has been derived in section 5.9 for the volume rate of flow under streamline conditions of a liquid through a uniform cylindrical tube, as the result of a pressure difference between its ends. The expression for Q, the volume of liquid flowing per second through a uniform tube of internal radius a and length l under a pressure difference p, is

$$Q = \frac{\pi p a^4}{8\eta l}$$

where η is the coefficient of viscosity. In deducing this equation it was assumed that the volume of liquid passing any section of the tube in a given time was constant. This is not true in the case of a gas, since the density depends strongly on pressure. In this case the flow through an element of tube of length dx must be considered, since the pressure difference between the ends of such an element is only dp. For such an element of the tube the volume rate of flow is q where

$$q = -\frac{\pi a^4}{8\eta} \frac{dp}{dx} \qquad (6.6)$$

The minus sign is inserted because the gas flow takes place in the direction of the negative pressure gradient.

If the flowing gas is at constant temperature, Boyle's Law holds and p/ρ is a constant, where ρ is the gas density at a pressure p. Thus, if p_1 and V_1 denote, respectively, the pressure at the entrance to the tube and the volume entering per second, and p_2 and V_2 denote the corresponding quantities at the exit of the tube

$$p_1 V_1 = p_2 V_2 = pq$$

Therefore,

$$p_1 V_1 = -\frac{\pi a^4 p}{8\eta}\frac{dp}{dx}$$

and hence

$$p_1 V_1 \int_0^l dx = -\frac{\pi a^4}{8\eta}\int_{p_1}^{p_2} p\,dp$$

so that

$$p_1 V_1 l = \frac{\pi a^4}{16\eta}(p_1^2 - p_2^2)$$

or

$$\eta = \frac{\pi a^4(p_1^2 - p_2^2)}{16 l p_1 V_1} \qquad (6.7)$$

This may be written

$$\eta = \frac{\pi a^4(p_1 - p_2)(p_1 + p_2)}{16 l p_1 V_1}$$

When $p_1 \simeq p_2$, the expression for η becomes

$$\eta = \pi^4 a(p_1 - p_2)/8l V_1 \qquad (6.8)$$

Equation (6.6) was deduced on the assumption that the fluid velocity at the walls of the tube is zero. This is true for a gas provided that the pressure exceeds a few millimetres of mercury, but is not the case at lower pressures.

As in the case of liquid flow, a kinetic-energy correction must be applied, and conditions should be arranged so that this correction is small.

Equation (6.7) no longer holds at gas pressures so low that the mean free path is comparable with the dimensions of the apparatus.

The gas flow that occurs under these latter conditions is termed 'molecular flow', and a gas at such pressures is often called a 'Knudsen' gas.

In a simple treatment of the flow of such a gas along a uniform tube, it is assumed that the only collisions the molecules make are with the wall of the tube, though in crossing the tube the molecules acquire a drift velocity parallel to the axis of the tube due to the pressure gradient along the tube. (In this way collisions with other gas molecules are introduced, though they are assumed not to affect significantly the progress of the molecules across the tube.) At the collisions with the wall it is assumed that all the drift velocity is lost, and that the directions of motion of the molecules on leaving the wall are quite random.

If the tube has a radius a, the average distance a molecule travels between successive collisions with the wall will be of the order of $2a$. (Some distances will be smaller and some larger, depending on the direction in which the molecule moves on leaving the wall.) If the mean velocity of the molecules is $\bar{c}$, the average time between successive collisions with the wall is $2a/\bar{c}$. During this time each molecule acquires a drift velocity in the direction of the axis of the tube due to the pressure gradient dp/dx along the tube, where x is the direction along the axis of the tube. In a length of tube dx the number of molecules is $n\pi a^2 dx$, where n is the number of molecules per unit volume, and, if each has a mass m, the mean acceleration suffered by each molecule in crossing the tube is $\pi a^2 dp/mn\pi a^2 dx = \frac{1}{\rho} dp/dx$, where ρ is the gas density. The acceleration continues for a time $2a/\bar{c}$, so that the velocity acquired by each molecule just before striking the wall is, on average, $\frac{1}{\rho}\frac{dp}{dx}\frac{2a}{\bar{c}}$, giving a mean drift velocity along the tube of $\frac{1}{\rho}\frac{dp}{dx}\frac{a}{\bar{c}}$

Since this acceleration process is occurring continuously, the mean volume of gas transported along the tube per second is

$$\pi a^2 \frac{a}{\rho\bar{c}}\frac{dp}{dx}$$

which is, if (p_1-p_2) is the pressure difference between the ends of a tube of length l,

$$Q = \frac{\pi a^3(p_1-p_2)}{\rho\bar{c}l} \tag{6.9}$$

It has been seen that at pressures above a few millimetres of mercury the volume rate of flow for a gas is given by equation (6.6), whereas at very low pressures it is given by equation (6.9). In the range of intermediate pressures the rate of flow must be given by an equation that tends to the form of (6.6) at high pressures and that of (6.9) at very low pressures. A satisfactory form is

$$Q = \frac{(p_1-p_2)a^3}{\rho\bar{c}l}\left(A+\frac{Ba}{L}\right)$$

where L is the mean free path and A and B are constants. This equation shows that in the design of pumping systems for use at low pressures it is important to have connections that are short and wide and that the width is more important than the length.

6.5 Measurements of gas viscosity

Gas viscosities are frequently obtained from measurements of the gas flow through capillary tubes. One such method is that due to Searle. When applied to air, the air is made to pass under streamline conditions, and at constant temperature, from a container of volume V through a capillary tube of internal radius a and length l into the atmosphere, using the arrangement shown in fig. 6.4.

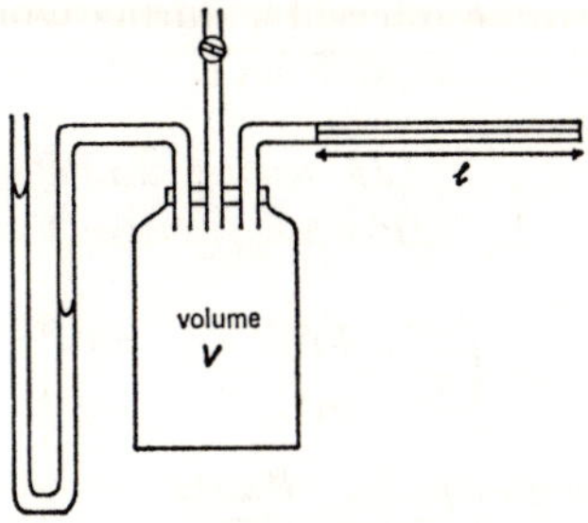

Figure 6.4

Differences of the gas pressure from atmospheric pressure are measured by a manometer containing a low density (and low vapour pressure) oil.

Let P be the atmospheric pressure and let the pressure of the air in the container fall from $P+p_1$ to $P+p_t$ in time τ. At some instant of time t let the pressure in the container be $P+p$. The p's are the pressure differences indicated by the manometer.

If V_1 is the volume of air entering the capillary per second at time t, and if the pressure changes from $P+p$ to $P+p+dp$ in time dt, then, if the temperature remains constant,

$$(P+p)\,V = (P+p+dp)(V+V_1dt)$$

or,
$$(P+p)\,V_1 = -V\frac{dp}{dt}$$

From the previous section

$$(P+p)\,V_1 = \frac{\pi a^4}{16\eta l}\,[(P+p)^2-P^2]$$

and hence
$$V\frac{dp}{dt} = -\frac{\pi a^4}{16\eta l}\,[(P+p+P)(P+p-P)]$$

$$= -\frac{\pi a^4 p(p+2P)}{16\eta l}$$

or,
$$\frac{dp}{dt} = Ap(2P+p)$$

where $$A = -\pi a^4/16\eta l V$$

so that $$\frac{dp}{p(2P+p)} = A\,dt$$

Resolve into partial fractions. Then,

$$\frac{dp}{p} - \frac{dp}{(2P+p)} = 2AP\,dt$$

Integrate over the time interval τ during which the pressure indicated by the manometer changes from p_i to p_f

$$\int_{p_i}^{p_f} \frac{dp}{p} - \int_{p_i}^{p_f} \frac{dp}{2P+p} = -\frac{2\pi a^4 P}{16\eta l V}\int_0^\tau dt$$

Then, $$\ln\left(\frac{p_f}{p_i}\right) - \ln\left(\frac{2P+p_f}{2P+p_i}\right) = -\frac{\pi a^4 P\tau}{8\eta l V}$$

or, $$\ln\left[\left(\frac{2P+p_f}{2P+p_i}\right)\frac{p_i}{p_f}\right] = \frac{P\pi a^4\tau}{8\eta l V} \qquad (6.10)$$

This treatment neglects the mass of gas that flows out of the manometer. One solution to this problem is to fit a manometer with a movable arm so that the manometer volume may be kept constant. However, the kinetic-energy correction is usually more important than the manometer correction.

The experiment described can be modified to measure the viscosity of air at different temperatures. In this case it is usual to change the temperature of the capillary tube only to, say, T_tK while that of the surroundings is T_sK. Then, the volume of air flowing through the capillary, though measured in terms of the temperature of the reservoir, must be corrected to that of the capillary, so that, using the ideal gas equation, the appropriate equation for the experiment becomes

$$\ln\left[\left(\frac{2P+p_f}{2P+p_i}\right)\frac{p_i}{p_f}\right] = \frac{P\pi a^4\tau T_s}{8\eta l V T_t} \qquad (6.11)$$

6.6 Results

The results of gas viscosity measurements show that typical values at room temperature are of the order of 10^{-5}N s m^{-2}. This may be compared with the value for water at the same temperature which is about 10^{-3}N s m^{-2}.

It has already been shown that on the basis of kinetic theory Maxwell predicted that the viscosity of a gas should be independent of pressure, a result which is found to hold over a wide range of pressures, and which was one of the early triumphs of the kinetic theory.

The viscosity of a gas increases with temperature, in marked contrast to the viscosities of liquids which, in general, decrease with temperature. Using the equation derived from the kinetic theory, namely,

$$\eta = \tfrac{1}{3}\rho\bar{c}L \qquad (6.5)$$

Maxwell predicted that η should be proportional to $T^{1/2}$, where T is the Kelvin temperature, for it can be shown that $\bar{c}$ is proportional to $T^{1/2}$ and equation (6.3) shows that ρ is proportional to $1/L$.

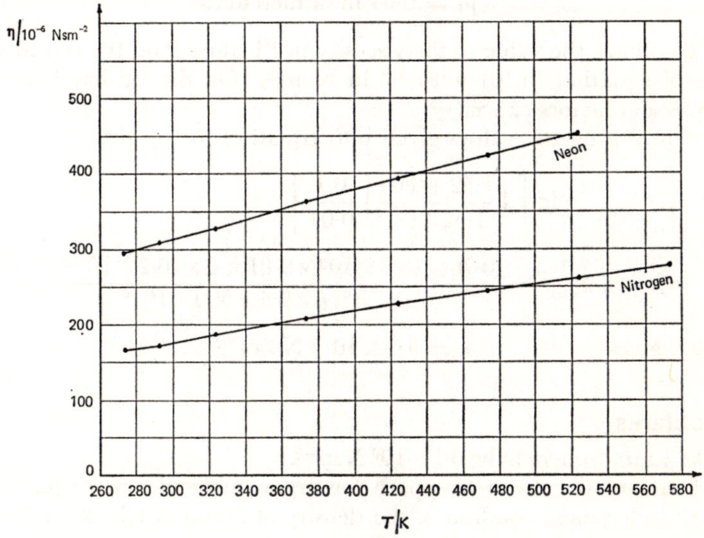

Figure 6.5

Sutherland has shown by a more rigorous approach that if η_T is the viscosity of a gas at T K and η_{273} is the viscosity at the ice point

$$\frac{\eta_T}{\eta_{273}} = \left(\frac{273+C}{T+C}\right)\left(\frac{T}{273}\right)^{3/2} \qquad (6.12)$$

C is a constant for a given gas and is usually referred to as Sutherland's constant. This formula agrees well with experiment, except at very high and very low temperatures.

The variation of viscosity with temperature for two gases is shown in fig. 6.5.

6.7 Example

A bulb of volume 500×10^{-6}m^3 has leading from it a uniform tube of length 0.4 m and internal radius $0{\cdot}20 \times 10^{-3}$ m. The bulb is filled with a gas at an initial pressure of 0·86 m of mercury and it is found that the pressure falls to 0·80 m of mercury in 30 seconds, the volume of the bulb remaining constant. Calculate the viscosity of the gas at the temperature of the measurements. Atmospheric pressure is 0·76 m of mercury and the density of mercury is $13{\cdot}6\times10^3$ kg m^{-3}.

Using the notation of equation (6.10)

$$p_i = 0{\cdot}10 \text{ m of mercury}$$
$$2P = 1{\cdot}52 \text{ m of mercury}$$
$$p_f = 0{\cdot}04 \text{ m of mercury.}$$

To obtain the value of the viscosity in SI units P on the right-hand side of equation (6.10) must be in N m^{-2}. (On the left-hand side all conversion factors cancel.)

Substituting the values given into equation (6.10) gives

$$\ln\left[\left(\frac{1{\cdot}52+0{\cdot}04}{1{\cdot}52+0{\cdot}1}\right)\frac{0{\cdot}1}{0{\cdot}04}\right] = \frac{0{\cdot}76\times13{\cdot}6\times10^3\times9{\cdot}81\times\pi\times(0{\cdot}2\times10^{-3})^4\times30}{8\times\eta\times0{\cdot}4\times500\times10^{-6}}$$

which gives $\eta = 1{\cdot}1\times10^{-5}$ N s m^{-2}

Problems

Take 1 atmosphere to be $0{\cdot}1\times10^6$ N m^{-2}.

1. Estimate the mean free path for argon molecules at s.t.p., given that, under these conditions, the density of argon is 1.78 kg m^{-3} and the viscosity is 21×10^{-6} N s m^{-2}.

2. A vertical capillary tube of length L and internal radius a_2 is connected at its upper end to another vertical tube of internal radius a_1. A pellet of mercury of length x is introduced into the top of the upper tube and takes t seconds to move through a vertical distance l in that tube. If η is the viscosity of air, ρ the density of mercury and P the atmospheric pressure in m of mercury, show that

$$\frac{l\,a_1^2}{t} = \frac{a_2^4}{16\eta L}\left\{\frac{[(P+x)\rho g]^2-[P\rho g]^2}{(P+x)\rho g}\right\}$$

Assume that $a_1 \gg a_2$, that there is no sticking between the mercury and the glass and neglect end effects. This is a version of Rankine's method for the determination of gas viscosities.

3. Raindrops of radius $5{\cdot}2\times10^{-6}$ m fall in air with a terminal velocity of 0·2 m min^{-1}. If the density of air is 1·3 kgm^{-3}, calculate the viscosity of air at the temperature of the measurement.

4. Calculate the gas kinetic diameter for hydrogen molecules if, at

s.t.p., the viscosity is $8 \cdot 6 \times 10^{-6}$ N s m^{-2}, the density is $89 \cdot 9 \times 10^{-3}$ kg m^{-3} and there are $2 \cdot 7 \times 10^{25}$ molecules per m^3.

Further reading

Christy, R. W. and Pytte, A., *The Structure of Matter*, W. A. Benjamin Inc., 1965, ch. 7, 8.

Jeans, J. H., *Introduction to the Kinetic Theory of Gases*, C.U.P., 1940.

CRANFIELD INSTITUTE OF TECHNOLOGY LIBRARY

7 surface effects

7.1 Introduction

In solids and liquids the basic particles are very close together and the bulk material shows short-range order. At the surface of a solid or liquid this order ends abruptly. The surface can thus be considered as a two-dimensional defect in the particle arrangement. The termination of the particle packing means that the particles in the surface layer have some bonds that cannot be satisfied because of the absence of neighbouring particles, and this will cause a change in the particle arrangement in the surface layer compared with that of the bulk material.

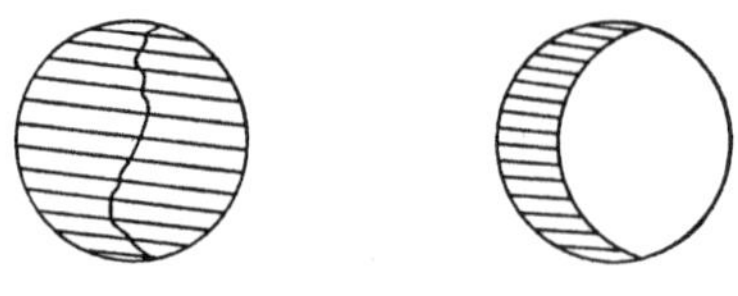

Figure 7.1

This suggests that the surface of a solid or liquid may exhibit properties not shown by the bulk material. Experiment shows many effects in which the expected behaviour is modified by the surface. For example, with some liquids it is possible to put more into a beaker than the geometrical volume, small quantities of water when placed on a dusty surface gather into nearly spherical drops, many liquids rise in a vertical capillary tube, and so on. One very important effect is the spontaneous contraction of the surface area of a liquid when the external forces change in a way that makes this possible. For example, if a soap film is formed on a wire frame across which a loose piece of cotton is tied, as in fig. 7.1a, when the film is broken on one side of the cotton, the cotton very rapidly becomes taut and takes the shape of an arc of a circle as in fig. 7.1b. In each of the examples quoted the liquid seems to be acted upon by forces additional to the gravitational force, and it is the object of this chapter to discuss these effects and interpret them in terms of intermolecular forces.

7.2 Free (or mechanical) surface energy

It has been shown that the intermolecular forces are repulsive at small intermolecular distances and attractive at larger ones, but are also of very short range, so that when two molecules are separated by more than a few molecular diameters the mutual force of attraction is effectively zero.

A molecule in the bulk of a liquid is able to move about to a considerable extent. The magnitude of the attractive forces exerted on it by its near neighbours will then vary with time, both in magnitude and direction, but the time average of these forces (considering times much larger than the period of molecular vibration) will be zero. In collisions the repulsive forces are brought into play. They may be regarded as changing the direction of motion of the liquid molecules in collisions. Since the attractive forces are very short range, the term bulk can be applied to any portion of a liquid (or solid) that is more than a few molecular diameters from the surface, say about 10^{-9} m away.

For any molecule that is closer to the surface than this distance the time average of the attractive forces is no longer zero, but has a finite resultant directed normally away from the surface, towards the bulk of the liquid. A sphere, known as the 'sphere of activity', may be drawn round each molecule, such that only those other molecules whose centres lie within the sphere exert any attraction on the molecule. Fig. 7.2 shows a section through a liquid normal to the surface. The circles indicate the sections of the spheres of activity in this plane. For molecule A, well within the bulk of the liquid, the resultant steady force is zero, which means that A can move through the liquid without doing work. On the other hand, for molecule B the sphere of activity intersects the surface, and there is thus a resultant steady attractive force R directed towards the bulk of the liquid as shown. The molecules of vapour above the surface are, except near the critical point, too few to have any appreciable effect.

As the molecules in the liquid surface are subject to an unbalanced attractive force inwards, they must, according to Newton's second law of motion, accelerate inwards. When this happens the number of molecules in the surface layer will decrease, so that the area of the surface will decrease. Complete collapse of the liquid is prevented because the repulsive forces ensure that no matter what changes in shape (and hence in surface area) occur, the liquid involved must occupy a definite volume. Thus, the surface area decreases to a minimum value, the nature of the minimum being governed by the externally imposed constraints on the liquid shape. A freely falling drop suffers no such con-

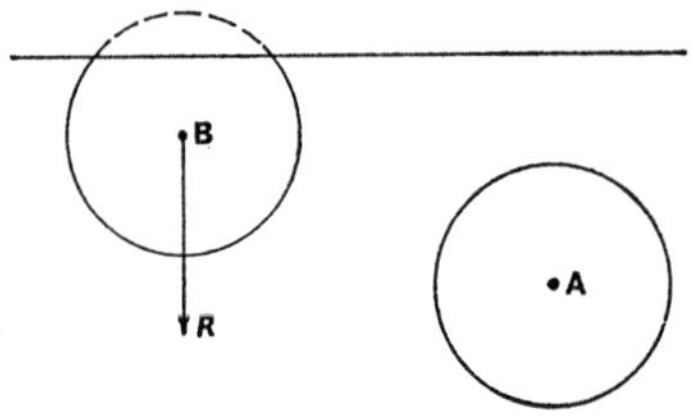

Figure 7.2

straint and thus takes up the shape that has the smallest surface area for a given volume, that is, a sphere. The repulsive forces serve to change the direction of motion of the molecules so that, once conditions are steady, as molecules leave the surface layer they are continually replaced by others from the bulk of the liquid.

A particle in the bulk of a solid occupies a mean position in which the resultant force exerted on it by its neighbours is zero. Those particles in the surface layers would experience an unbalanced attractive force towards the bulk of the solid if the interparticle spacing in the surface layer had the same value as in the bulk. This attractive force would cause the surface particles to move inwards, until the repulsive force brought into play by reducing the interparticle spacing exactly compensated the unbalanced attractive force, so that the surface particles were again in equilibrium positions, but with a reduced separation. Clearly this argument can be extended to cover partially crystalline and amorphous solids, since in all cases there must be a slight particle rearrangement in the surface layers.

When the surface area of a liquid is increased, molecules must be brought from the bulk of the liquid to achieve this. Once the sphere of activity of a molecule being brought towards the surface cuts the surface, the molecule is subject to a steady attractive force directed towards the bulk of the liquid and normal to the surface. This force increases as the distance from the surface decreases. To move the molecule work must be done against this steady force, so that the molecules in the surface layer have a greater potential energy than molecules in the bulk of the liquid. Hence, this force directed towards the bulk of the liquid shows itself as a force opposing the increase of the surface area of the liquid. It has already been mentioned that, except when the temperature is close to the critical temperature, the effect of the vapour is negligible. In the surface layer the molecule has a number of chemical bonds permanently broken because it is not completely surrounded by other molecules. The extra potential energy of the surface layer corresponds to that needed to break a certain number of bonds normally formed in the bulk of the liquid. Let u be the energy needed to break one intermolecular bond, that is, the energy needed to separate a molecule from one of its nearest neighbours. If, in the bulk of the liquid the number of nearest neighbours is m, the molecule is bound in the liquid by an energy mu. When the molecule is placed in the surface of the liquid the number of nearest neighbours becomes n, where n is less than m, so that the molecule is now bound by an energy nu. Thus the molecule acquires energy $(m-n)u$ by virtue of being in the surface.

These arguments can be applied equally well to solids, though the process of taking a molecule from the bulk of the material and placing it in the surface to increase the surface area may be difficult to visualize, indicating that the surface of a solid, like that of a liquid, is a seat of potential energy.

To enlarge the surface area of a liquid or solid, work must be done against the steady attractive forces (directed towards the bulk of the material) found in the surface layer, since the number of molecules in the surface must be increased. The work that must be done under isothermal conditions to increase the surface area by unity is called the free surface energy, γ. It has units in the SI system J m^{-2}. Thus, γ is the mechanical work that can be recovered when the surface area decreases isothermally by unity. If a value of γ is specified with reference to at particular material in a particular phase, for example, γ for water at room temperature is about 72×10^{-3} J m^{-2}, it is assumed that the material, in this case water, is in contact only with its vapour (the atmosphere has very little effect). Clearly free surface energy can be defined for any interface, that is, for any region in which the environment of the molecule differs from that in the bulk material. When the environment is not the vapour of the material, the second medium must be specified, for example, the free surface energy of the water–benzene interface at room temperature is about 35×10^{-3} J m^{-2}.

Consider a horizontal wire frame of the shape shown in fig. 7.3, fitted with an arm XX that can move without friction. Let a liquid film be formed as shown by the shaded area, and let it be in equilibrium under a force F. Let the arm be moved slowly (so that all changes may be isothermal) through a distance δx in the direction of F. The surface area is increased by $2\ (XX)\ \delta x$ (the factor 2 arises because the film has two surfaces) and thus the free surface energy is increased by an amount $2\ (XX)\ \delta x\gamma$, where γ is the free surface energy of the liquid (in contact with its vapour). The applied force F does work on the liquid film equal to $-F\delta x$ (the negative sign is used since work done by a system is counted positive). The principle of virtual work states that when a small displacement occurs of a system in equilibrium the work done is zero.

Thus $$-F\delta x + 2(XX)\delta x\gamma = 0$$

or $$F = 2\gamma(XX)$$

This may be interpreted as showing that a liquid surface exerts a mechanical force on any solid surface with which it is in contact, the magnitude of this force per unit length being numerically equal to the free surface energy of the free liquid surface.

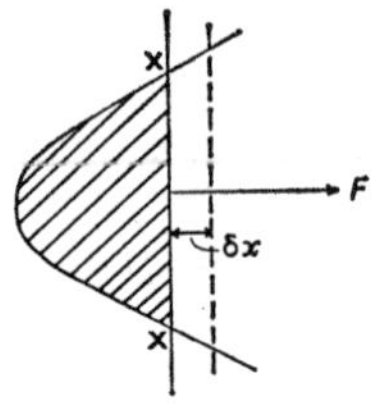

Figure 7.3

7.3 Total surface energy

The total change in energy of the surface is not, in general, equal to the change in free surface energy. Basically, this is because the free surface energy decreases as the temperature rises. If the surface is enlarged isothermally, heat flows in to the material from the surroundings. Let a liquid surface of area A be enlarged by an amount dA in a reversible manner. The work done on the surface is $dW = -\gamma dA$. Hence, the First Law of Thermodynamics applied to this process becomes

$$dQ = dU + dW = dU - \gamma dA \tag{7·1}$$

where dQ is the heat transfer between the liquid and its surroundings and dU is the change in the internal energy of the liquid. Since γ does not depend on A the first TdS equation for a liquid surface can be written (see Zemansky, Heat and Thermodynamics)

$$dQ = TdS = C_A dT - T\frac{d\gamma}{dT}dA$$

where C_A is the heat capacity of the liquid at constant surface area. In an isothermal process dT is zero and then

$$dQ = -T\frac{d\gamma}{dT}dA$$

Substituting in equation (7.1) gives

$$-T\frac{d\gamma}{dT}dA = dU - \gamma dA$$

or,
$$dU = \left(\gamma - T\frac{d\gamma}{dT}\right)dA$$

In a finite change from surface area A_i to A_f, with a corresponding change in internal energy from U_i to U_f,

$$\frac{U_f - U_i}{A_f - A_i} = \gamma - T\frac{d\gamma}{dT} = \lambda. \qquad (7.2)$$

λ, which has the same units as γ, is interpreted as the energy associated with unit area of surface, that is, the total surface energy per unit area.

7.4 Young's relation and liquid spreading

It has been shown in section 7.2 that the molecular arrangement in the surface of a solid is not the same as that in the bulk material, giving rise to the concept of a free surface energy for the solid. If the surface is now covered by a liquid, the surface energy of the solid is reduced, because the one-sidedness of the environment of the surface molecules is reduced. The new value of the free surface energy depends on the nature of both the solid and liquid, and is referred to as the 'free interfacial energy' for the two media concerned. When a liquid is placed on part of a solid surface it may spread over the surface and so reduce the surface energy of the solid. As it does this it enlarges its own surface, thus increasing its own surface energy. Spreading will continue as long as the sum of the free surface energies is reduced as a result, and will cease when the sum of the free surface energies is a minimum.

Let the free surface energy of the solid-vapour interface be γ_{SV}, of the liquid-vapour interface be γ_{LV} and of the liquid-solid interface be γ_{LS}. The liquid will spread over the (perfectly uniform) solid surface completely if the increase in free surface energy resulting from the increase in area of the liquid-vapour interface and liquid-solid interface is less than the decrease due to the reduction in area of the solid-vapour interface, neglecting any gravitational energy changes, that is, if

$$\gamma_{SV} > \gamma_{LS} + \gamma_{LV} \qquad (7.3)$$

This also assumes that the liquid film is sufficiently thick so that γ_{LS} and γ_{LV} act independently.

In the case of a drop of a liquid A resting on a liquid B with which it does not mix, spreading will occur if there is a reduction in the sum of the free surface energies, that is, if

$$\gamma_A + \gamma_{AB} < \gamma_B$$

that is, the spreading coefficient S, defined as

$$S = \gamma_B - \gamma_A - \gamma_{AB} \qquad (7.4)$$

should be positive. This will be the case if

$$\gamma_A - \gamma_B + \gamma_{AB} < 0$$

or

$$\gamma_A - \gamma_B - \gamma_{AB} < 0.$$

Again it has been assumed that the liquid that spreads does so in a thick (or duplex) film so that γ_A and γ_{AB} act independently.

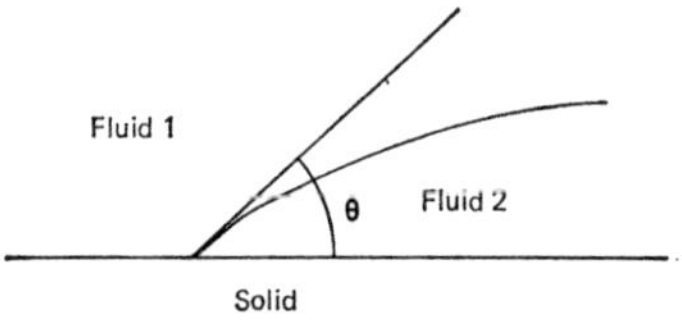

Figure 7.4

Two immiscible liquids in contact have an interface with a free surface energy γ_{AB}. When the liquids are separated, two surfaces with free surface energies γ_A and γ_B respectively are formed, so that the change in free surface energy is $\gamma_A + \gamma_B - \gamma_{AB}$. This quantity, which represents the work done in separating two liquids in contact over unit area, is called the work of adhesion, W_{AB}, of the two liquids. If a column of liquid A having unit area of cross-section is separated into two parts the work done is $2\gamma_A$, since two surfaces, each of unit area, are formed. This quantity is called the work of cohesion, W_A, for liquid A. The spreading coefficient can then be written $S = W_{AB} - W_A$, so that spreading is expected if $W_{AB} > W_A$, that is, if the adhesion of A for B is greater than the cohesion of A.

Experimentally it is found that where liquid A spreads on liquid B in the form of a duplex film it later collects into large drops in equilibrium with a monolayer film on liquid B.

When two immiscible fluids meet at a solid surface, and are in an equilibrium configuration, the surface of separation of the fluids makes a definite angle with the solid surface. Fig. 7 .4 shows a possible situation where the solid surface is, in fact, plane. If one of the fluids is a gas, it is usual to refer to the angle between the solid and the tangent to the liquid surface at the line of contact with the solid, measured in the liquid, as the angle of contact for that particular combination of solid and liquid. If, in fig. 7.4, fluid 2 is liquid and fluid 1 is gas, then θ is the angle of contact for the particular solid-liquid combination.

Consider a solid with a plane surface of infinite extent, covered by a liquid which is in equilibrium and stretches to infinity over half of the solid surface, as shown in fig. 7.5. The potential energy of this system will not be changed if the line of contact between the liquid-vapour interface and the solid moves through a small distance δx to the left or right in the diagram. Let the movement be to the left, so that the new position is shown dotted. Considering unit length of the interface, the area of the solid-vapour interface decreases by δx, that of the solid-liquid interface increases by δx and that of the liquid-vapour interface increases by $\delta x \cos\theta$, where θ is the angle of contact for the liquid and solid concerned.

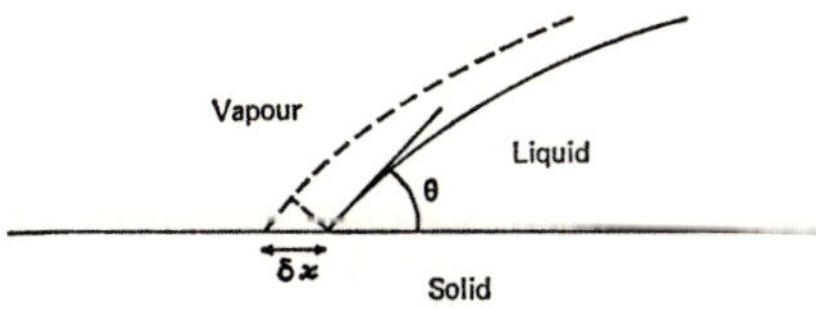

Figure 7.5

By the principle of virtual work the total potential energy change is zero.

Thus, $$-\gamma_{SV}\delta x+\gamma_{LS}\delta x+\gamma_{LV}\delta x\cos\theta = 0$$

or $$\cos\theta = \frac{\gamma_{SV}-\gamma_{LS}}{\gamma_{LV}} \qquad (7.5)$$

Equation (7.5) is known as Young's relation.

For a given combination of liquid and solid the value of the angle of contact depends to some extent on whether the liquid is advancing or receding over the surface. Advancing angles of contact are usually larger than receding ones, as shown in fig. 7.6.

Fig. 7.6

When, for a given solid-liquid pair, θ is greater than 90° the liquid is said to be 'non-wetting' with respect to that solid; when $90° > \theta > 0°$ the liquid is said to be 'partially wetting' and when $\theta = 0°$ it is said to 'wet' the surface. In practice θ is only zero if the liquid has, in fact, previously wetted the surface, so that a very thin film of liquid remains on the surface, as in fig. 7.7.

The angle of contact is the only directly measurable quantity which gives any information about interfacial energies. Its value is extremely sensitive to surface contamination. One method of determining θ for a given solid-liquid interface is to take a plane plate of the solid and partially immerse it in the liquid, as in fig. 7.8a. The plate is then rotated about an axis lying in the plate, parallel to the liquid surface, until the liquid meets the solid on one surface without curvature of the liquid surface, as in fig. 7.8b. θ is then easily measured.

Now that Young's relation has been obtained, the condition that a liquid spreads over a solid can be considered in terms of the angle of contact. Equation (7.3) shows that spontaneous spreading will occur if

$$\gamma_{SV} > \gamma_{LS} + \gamma_{LV}$$

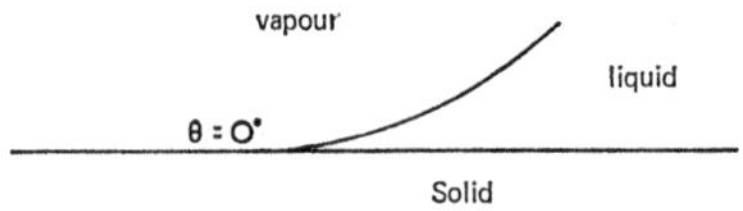

Figure 7.7

that is, if the adhesion of the liquid for the solid, $\gamma_{SV}+\gamma_{LV}-\gamma_{LS}$ exceeds the cohesion of the liquid $2\gamma_{LV}$. Using equation (7.5) the work of adhesion of liquid to solid becomes

$$\gamma_{LV}(1+\cos\theta)$$

so that the condition for the spreading of the liquid over the solid becomes

$$\gamma_{LV}(1+\cos\theta) > 2\gamma_{LV} \tag{7.6}$$

But, as long as θ is finite, $2\gamma_{LV} > \gamma_{LV}(1+\cos\theta)$, so that, for a liquid to spread over a solid, θ must be equal to zero. This means that the work of adhesion of the liquid to the solid must be as great as the work of cohesion of the liquid. It is possible for the work of adhesion to be greater than the work of cohesion, but this cannot be expressed in terms of θ.

When the liquid does not spread to cover the complete surface the effect of changes of gravitational potential energy must be considered.

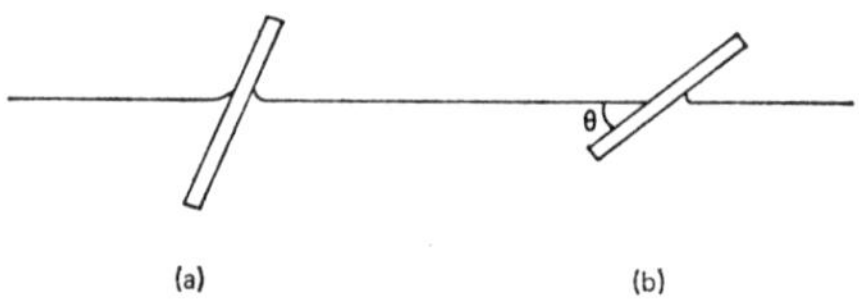

Figure 7.8

7.5 Intrinsic pressure

In the surface region of a liquid where the molecules are subjected to an unbalanced steady attractive force, there is a gradual build-up of pressure toward the bulk of the liquid which is quite independent of gravitational forces. Consider a liquid in equilibrium, having a plane surface; a section normal to the surface is shown in fig. 7.9. AA is the surface and l is the radius of the sphere of activity. Consider a very narrow layer of liquid a distance x from the free surface and bounded by the surfaces BB and CC. The force acting on BB as a result of the unbalanced steady force is greater than that acting on CC, since BB is closer to the free surface than CC. Hence, to maintain the state of equilibrium, the pressure exerted on CC by the liquid below the element must be greater than that exerted on BB by the liquid above the element. This reasoning may be applied to any elementary layer within

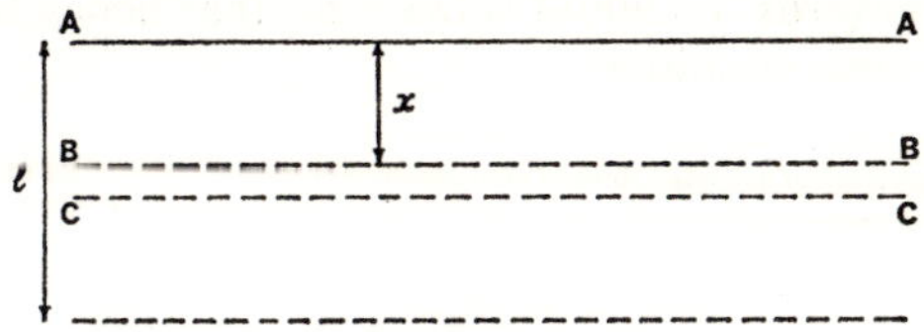

Figure 7.9

a distance l of the free surface, and hence the pressure exerted by the liquid arising from this mechanism must increase with distance from the free surface over the anomalous surface region of thickness l, and thereafter remain constant. This pressure is referred to as the 'intrinsic pressure' and should not be confused with the hydrostatic pressure, which arises from the effect of gravity.

When the free surface is curved and is concave towards the bulk of the liquid, as in fig. 7.10 (the curvature is drawn much exaggerated), the intrinsic pressure is greater than in the case where the surface is plane, the difference between the intrinsic pressure in the two cases

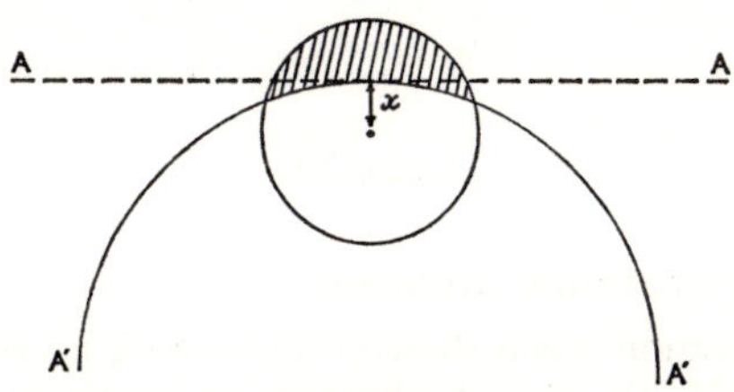

Figure 7.10

being the 'excess pressure'. A molecule that is a distance x below the free surface A'A' ($x<l$) suffers an inwards force proportional to the volume of the sphere of activity shown shaded, since the steady force due to molecules in an equal volume of the sphere activity below the surface is now unbalanced. Had the surface been plane, however (AA in fig. 7.10), the molecules giving rise to the unbalanced inwards force would occupy a volume equal to the shaded volume of the sphere of activity that lies above AA. This is smaller than the total shaded volume, and thus the steady inwards force would be smaller. Hence, by applying the arguments of the above paragraph, it can be seen that the pressure beneath a liquid surface that is concave towards the liquid

increases with distance from the surface, becomes constant at a distance equal to the radius of the sphere of activity, and has a greater value than that below a plane surface under the same conditions. This pressure is independent of any gravitational effects and arises solely from intermolecular forces. Similarly, the pressure beneath a surface that is convex towards the liquid is less than that beneath a plane surface under the same conditions.

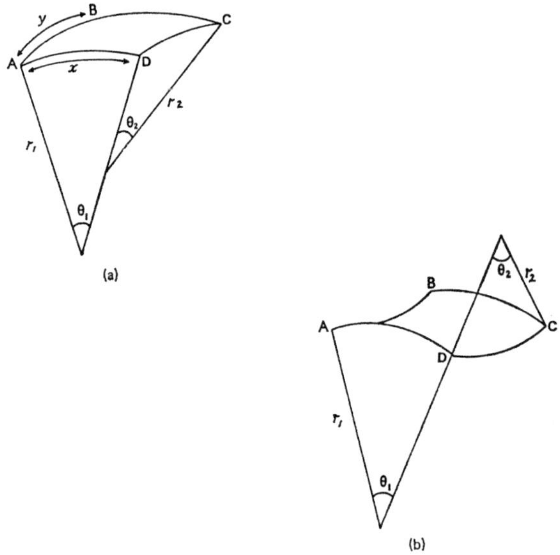

Figure 7.11

7.6 The excess pressure theorem

In the previous section it was shown that, in a liquid with a surface that is concave towards the liquid, there is an excess pressure that has a constant value throughout the whole of the liquid (except for a very narrow surface region). The intrinsic pressure in a liquid may be considered as the sum of two parts: the value that would be found if the surface were plane, plus the excess or defect due to the curvature of the surface. Since the former part always cancels out from any equations, only the latter will be considered explicitly in future.

Any region of a surface is specified by the values of the two principal radii of curvature in that region. Consider a small rectangular element of a free liquid surface having principal radii of curvature r_1 and r_2. These curvatures may have the same sign as in fig. 7.11a or be of opposite sign as in fig. 7.11b. For clarity in drawing, the analysis will be carried out for the case shown in fig. 7.11a. Let the pressure on the concave side of surface be p_1 and that on the convex side be p_2. It has been shown that $p_1 > p_2$.

Take the x axis parallel to AD and the y axis parallel to AB. Since the chosen rectangle is very small, AD and AB are effectively straight and the area ABCD is effectively plane and of magnitude AD.AB $= xy = \alpha$. Let the surface area be expanded isothermally by an infinitesimal amount by allowing the radii of curvature to increase by δr, the centres of curvature remaining stationary. The change in area, $\delta \alpha$, is given by

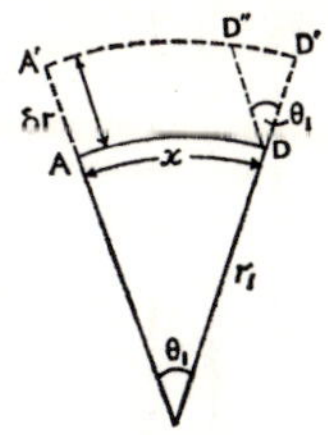

Figure 7.12

$$\delta\alpha = \delta(xy) = x\delta y + y\delta x$$

Fig. 7.12 shows A^{I} D^{I}, the new position of AD after the expansion or the surface, and has DD^{II} drawn parallel to AA^{I}, making A^{I} D^{II} equal to AD, and D^{II} D^{I} equal to δx.

Then $$\theta_1 = \frac{\delta x}{\delta r} = \frac{x}{r_1}$$

so that $$\delta x = \frac{x}{r_1}\,\delta r$$

Similarly, $$\delta y = \theta_2 \delta = \frac{y}{r_2}\,\delta r$$

Hence, $$\delta\alpha = \frac{xy\delta r}{r_1} + \frac{xy\delta r}{r_2}$$

The work that must be done on the liquid to produce the increase in area is $-(p_1-p_2)xy\delta r = -pxy\delta r$, where $p = p_1-p_2$. The increase in free surface energy is $\gamma\delta\alpha$. Therefore, applying the principle of virtual work to the element of surface,

$$-pxy\delta r + \gamma\delta\alpha = 0$$

or, $$p = \gamma\left(\frac{1}{r_1} + \frac{1}{r_2}\right) \qquad (7.7)$$

This gives the pressure difference that occurs on crossing any liquid surface. Clearly the result holds for any fluid-fluid interface provided the appropriate free interfacial energy is used for γ. When the surface is plane both r_1 and r_2 are infinite and p is zero. If a liquid film is being considered, two pressure differences occur of the same sign and magnitude, so that the pressure difference between the outer and inner surfaces of the film is

$$p = 2\gamma\left(\frac{1}{r_1} + \frac{1}{r_2}\right) \tag{7.8}$$

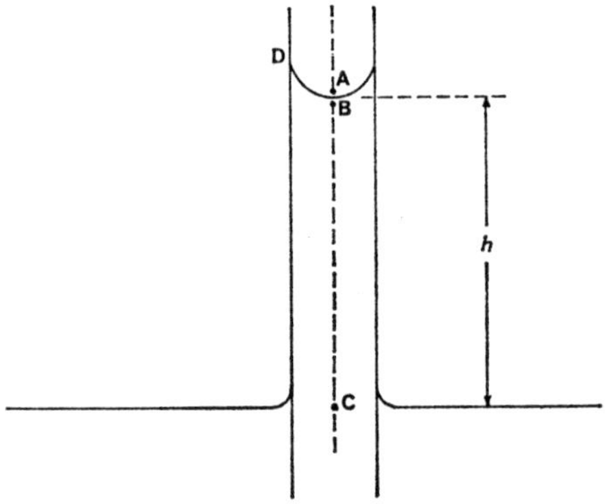

Figure 7.13

In the case of a spherical drop of liquid where $r_1 = r_2 = r$

$$p = \frac{2\gamma}{r}$$

and in the case of a spherical bubble

$$p = \frac{4\gamma}{r}$$

7.7 Capillary rise

One of the most striking surface effects is the rise (sometimes fall) of a liquid in a vertical tube partly immersed in it, to a height above (or below) that predicted by hydrostatic principles. A vertical, uniform cylindrical tube is shown in fig. 7.13, dipping into a liquid of free surface energy γ. Let the liquid rise in this case to an equilibrium position in which the lowest part of the meniscus in the tube is at a height h above the level of the liquid outside the tube and well away from the tube. (Experiment shows that liquids for which the angle of contact with the material of the tube is less than 90° rise in the tube and those for which it is greater than 90° are depressed.) The meniscus in the tube will be a surface that is symmetrical about the axis of the tube, and on the axis of the tube the two principal radii of curvature will be equal. Let their value be R. At a point A, just above the surface of the liquid the pressure p_A is atmospheric. On moving to a point B, just below the surface of the liquid and on the axis of the tube, the pressure decreases by an amount $2\gamma/R$, so that $p_B = p_A - 2\gamma/R$. At a point C, which is at the level of the liquid surface outside the tube, the pressure must again be atmospheric to maintain equilibrium. Hence $p_C = p_A$. But p_C is also equal to $p_B + \rho g h$, where ρ is the density of the liquid at the temperature of the experiment and g is the intensity of gravity. Thus

$$p_C = p_B + \rho g h = p_A - \frac{2\gamma}{R} + \rho g h$$

so that

$$h = \frac{2\gamma}{\rho g R} \tag{7.9}$$

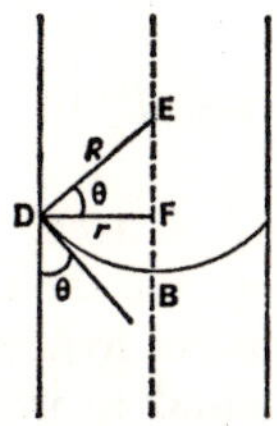

Figure 7.14

This gives h in terms of R, the radius of curvature at the lowest point of the meniscus. This relationship is valid whatever the diameter of the tube or the angle of contact. Unfortunately R cannot be measured directly.

If the tube is narrow the height BD is small and hence the change in pressure in going from B to D in the liquid is small. Thus p in equation (7.7) remains practically constant and hence so does $\left(\frac{1}{r_1} + \frac{1}{r_2}\right)$. If it is assumed that r_1 and r_2 individually remain constant, then the meniscus may be considered as part of a spherical surface. An enlarged view of this is shown in fig. 7.14. The tangent to the liquid surface along the line of contact with the solid defines the angle of contact θ. The normal to this tangent along the line of solid-liquid contact intersects the axis of the tube in the point E, and DE is R the radius of curvature of the meniscus. DF is the radius of the tube r.

Now $\frac{DF}{DE} = \cos\theta$ so that $R = \frac{r}{\cos\theta}$ and $h = \frac{2\gamma\cos\theta}{\rho g r}$ (7.10)

This equation is a very good approximation when r is small. When θ is greater than 90°, $\cos\theta$ is negative and thus h is negative. The liquid is then depressed in the tube as in fig. 7.15. If the liquid wets the material of the tube, θ is zero, $\cos\theta$ equals unity and

$$h = \frac{2\gamma}{\rho g r} \tag{7.11}$$

On this viewpoint, the liquid rises in the tube because a meniscus is formed when the tube is dipped into the liquid. This meniscus is curved, since the liquid surface must meet the walls at the angle of contact yet be horizontal on the axis of the tube, and hence there is a pressure drop in passing through the liquid surface of amount $2\gamma/R$.

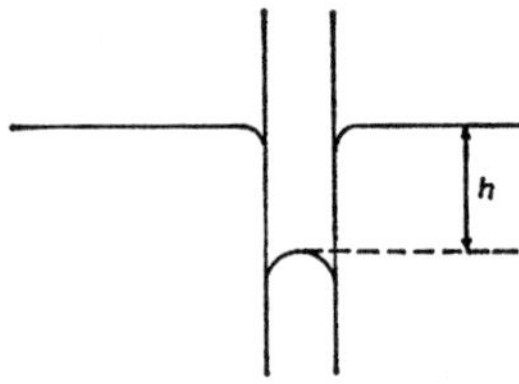

Figure 7.15

This pressure difference is unbalanced, and hence the atmospheric pressure acting on the liquid outside the tube pushes the liquid up the tube, until the resulting increase in hydrostatic pressure at the bottom of the liquid column is just equal to the drop in pressure found on passing through the meniscus. If this experiment is performed, the liquid is seen to rise gradually to an equilibrium position. In the absence of friction the liquid would be expected to acquire momentum as a result of the force acting on it, which would cause it to overshoot the equilibrium position so that oscillations would be set up. This is not observed, so the kinetic energy must be dissipated against viscous forces in the liquid.

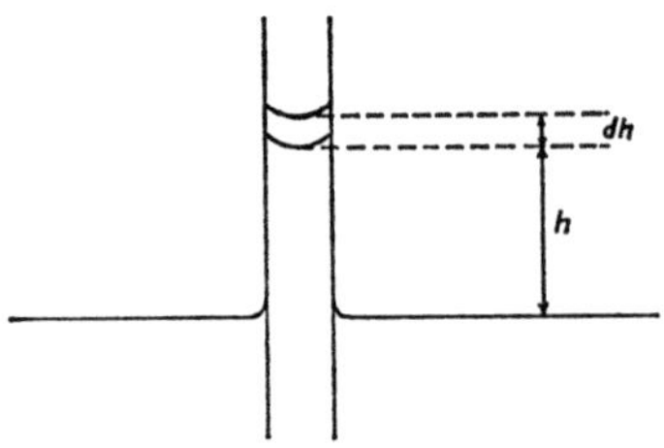

Figure 7.16

It is of interest to approach this problem from a consideration of free surface energies. The fact that the liquid rises in the tube indicates a reduction in the free surface energy of the system, and this has a minimum value when equilibrium is reached. Consider a combination of tube-material and liquid in which the liquid rises. As the liquid rises, the liquid-solid interface increases in area, but the area of the solid-vapour interface decreases. In a uniform tube the area of the liquid-vapour interface is constant. At the same time the potential energy of the liquid column is increased. The liquid continues to rise so long as the increase in potential energy of the liquid is less than the decrease in free surface energy of the system. As the liquid rises solid molecules at the surface become covered with liquid. The inward attraction towards the bulk of the solid is reduced, that is, their free surface energy is reduced and they more closely resemble molecules in the bulk solid.

When equilibrium is reached application of the principle of virtual work means that the change in energy for an infinitesimal displacement of the liquid surface is zero. Consider the situation shown in fig. 7.16, in which a liquid of density ρ, which only partially wets the solid, rises in a tube of radius r to an equilibrium height h. Let the height of the liquid column be increased by dh. The change in free surface energy is $2\pi r(\gamma_{LS}-\gamma_{SV})\ dh$ and the change in potential energy of the liquid is $\pi r^2 h\rho g dh$.

Since the system is in equilibrium, the total change in energy is zero.

Thus, $2\pi r(\gamma_{LS}-\gamma_{SV})\,dh+\pi r^2 h\rho g\,dh = 0$

and
$$h = \frac{2(\gamma_{SV}-\gamma_{LS})}{r\rho g} \tag{7.12}$$

Now $\gamma_{SV}-\gamma_{LS} = \gamma_{LV}\cos\theta$ (Young's relation),

and thus
$$h = \frac{2\gamma\cos\theta}{r\rho g} \tag{7.10}$$

In calculating the increase of the potential energy of the liquid column the liquid in the meniscus was neglected. The work done against gravity should have included the term

$$\left(\begin{array}{l}\text{weight of liquid above the horizontal}\\ \text{tangent plane to the meniscus}\end{array}\right) dh.$$

In the case of a liquid for which the angle of contact with the material of the tube is zero, and the bore of the tube is sufficiently narrow for the meniscus to be treated as a hemisphere, the weight of the liquid in the meniscus is

$$(\pi r^3-\tfrac{2}{3}\pi r^3)\rho g = \tfrac{1}{3}\pi r^3\rho g$$

Thus, for the complete liquid column, the work done against gravity is

$$(\pi r^2 h\rho g+\tfrac{1}{3}\pi r^3\rho g)\,dh$$

The change in free surface energy is still $2\pi r(\gamma_{LS}-\gamma_{SV})\,dh$, giving

$$\pi r^2\rho g\left(h+\frac{r}{3}\right) dh+2\pi r(\gamma_{LS}-\gamma_{SV})\,dh = 0$$

Young's relation in this case is $\gamma_{SV}-\gamma_{LS} = \gamma_{LV}$ and hence

$$\gamma = \frac{r\rho g}{2}\left(h+\frac{r}{3}\right) \tag{7.13}$$

Equation (7.13) is satisfactory when r is much less than h, but when this is not so it no longer gives an accurate result. One approach in such cases is to consider the principal section of the meniscus to be half an ellipsoid. Then it can be shown that, in the case of a liquid that wets the solid, γ is given by

$$\gamma = \frac{r\rho g h}{2}\left[1+\frac{r}{3h}+0{\cdot}111\frac{r^2}{h^2}+0{\cdot}0741\frac{r^3}{h^3}\right]$$

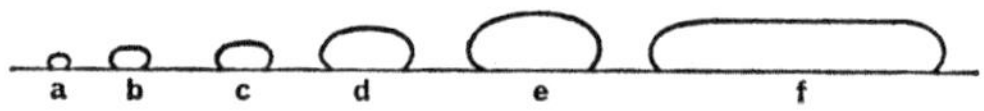

Figure 7.17

7.8 The sessile drop

If a very small volume of a liquid is placed on a plane horizontal surface of a material that it does not wet, the liquid collects into a shape that is almost spherical. The effect of the intermolecular forces in trying to make the surface area a minimum is much more important than that of the gravitational forces in trying to lower the centre of gravity. As more liquid is added to the drop it grows larger, but becomes slightly flattened as the effect of the gravitational forces becomes more important. With the addition of further liquid the drop grows progressively larger, but becomes more flattened. The depth increases until a certain value is reached. Then adding more liquid causes the depth to decrease slightly, and it reaches a steady value with the surface of the drop flat. The changes in shape that occur as more liquid is added to the drop are shown in fig. 7.17.

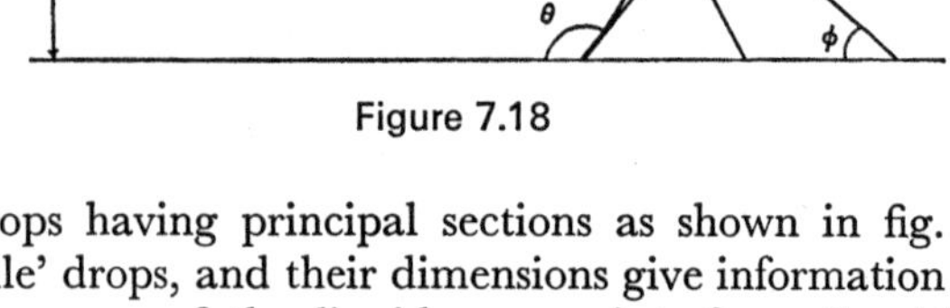

Figure 7.18

Large drops having principal sections as shown in fig. 7.17f are termed 'sessile' drops, and their dimensions give information about the free surface energy of the liquid-vapour interface. Consider a very large drop of, say, 1 metre diameter, a principal section of which is shown in fig. 7.18. Any small part of the circumference in a plane at right angles to the plane of the paper may be taken as straight, so that the only curvatures in the plane of the paper, that is, in a vertical plane, need be considered. Let s be the distance along the profile of the drop in a given vertical plane, measured from some arbitrary point 0 on the horizontal surface of the drop. At a point P on the profile the local radius of curvature of the surface is r, and the line joining P to the local centre of curvature makes an angle ϕ with the vertical. The value of r is given by $\partial s/\partial\phi$. The pressure just above the liquid surface at P is atmospheric; that just below the surface is atmospheric plus γ/r. But this is also equal to atmospheric pressure plus $y\rho g$, where y is the vertical distance from the horizontal surface to P and ρ is the density of the liquid.

Thus,
$$y\rho g = \gamma\frac{\partial\phi}{\partial s}$$

that is,
$$\gamma\frac{\partial\phi}{\partial y}\frac{\partial y}{\partial s} = y\rho g$$

Now
$$\frac{\partial y}{\partial s} = \sin\phi$$

Hence
$$\gamma\frac{\partial\phi}{\partial y}\sin\phi = y\rho g$$

or, separating the variables,
$$\sin\phi d\phi = \frac{gy\rho}{\gamma}dy$$

Integrate from the top of the liquid drop where ϕ and y are both zero to the point of maximum bulge where y equals h and ϕ is $\pi/2$.

Thus
$$\Big[-\cos\phi\Big]_{0}^{\pi/2} = \frac{\rho g}{2\gamma}\Big[y^2\Big]_{0}^{h}$$

or
$$1 = \frac{\rho g h^2}{2\gamma}$$

that is,
$$\gamma = \frac{\rho g h^2}{2} \tag{7.14}$$

a result that is independent of the angle of contact. If the integration is performed over the range $y = 0$ to $y = H$, the total height of the drop ϕ varies from zero when y is zero to θ, the angle of contact, when y equals H.

Thus
$$1-\cos\theta = \frac{\rho g H^2}{2\gamma} \tag{7.15}$$

7.9 The variation of free surface energy with temperature

For all liquids the free surface energy of the liquid-vapour interface decreases with increase of temperature, and becomes zero in the neighbourhood of the critical temperature. The variation of the free surface energy of water over the temperature range 0–100°C is shown in fig. 7.19. For a pure liquid in equilibrium with its vapour the free surface

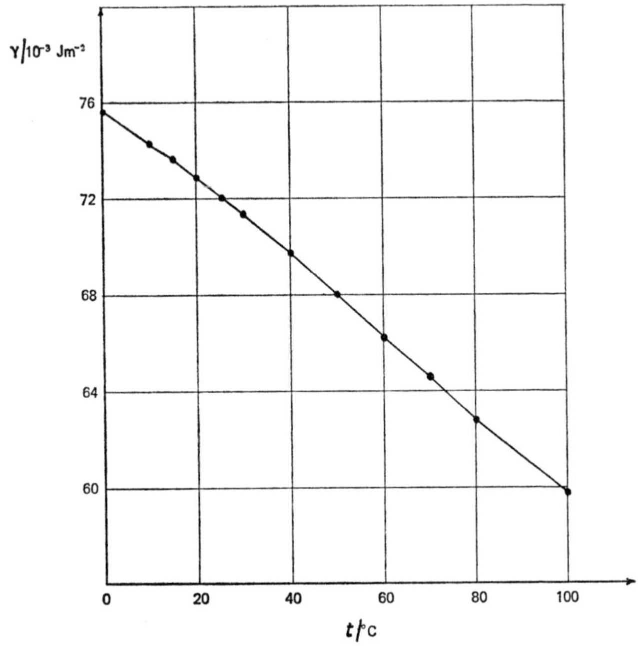

Figure 7.19

energy at $t°C$, γ_t, is related to the value at $0°C$, γ_0, by an equation of the type

$$\gamma_t = \gamma_0 \left(1 - \frac{t}{t^{\text{I}}}\right)^n \qquad (7.16)$$

where $t^{\text{I}}\,°C$ is a temperature a few degrees below the critical temperature and n is a constant having a value between 1 and 2.

7.10 The measurement of free surface energy for liquid-vapour interfaces

Any effect for which γ appears in the equations describing it can be used for the measurement of γ, but only a few such methods give reliable results; some of these will now be considered.

Methods for the determination of γ in which the liquid-vapour interface is continually renewed are termed 'dynamic' methods; those in which there is no such renewal are 'static'.

(*a*) *Capillary rise method*

This is a static method. To use equation (7.10), $2\gamma\cos\theta = \rho grh$, the tube must have a very narrow bore. However, narrow-bore tubes are very difficult to clean. Also it is difficult to make sure that the cross-section is circular at the position of the meniscus and to measure the radius accurately. The method is easily adapted for use in a vacuum or at high pressures and, by using a U-tube, for investigating the variation of γ with temperature.

In very precise work, ρ in equation (7.10) must be replaced by the difference in density of the gas and liquid present, since the pressure at C in fig. 7.13 will be greater than that at A by the pressure due to the column of vapour of height h.

Figure 7.20

When only a very small quantity of liquid is available, a modification of the above method due to Ferguson may be used. A drop of liquid is placed in horizontal capillary tube, one end of which is ground flat, as in fig. 7.20. The other end of the tube is connected to a source of gas pressure which is increased until the liquid is forced along to the plane end of the tube and its surface at this end of the tube becomes plane. The excess pressure over that of the atmosphere needed to achieve this is $2\gamma/r$, where r is the radius of the liquid surface, assumed spherical. For a liquid that wets the tube r is also the radius of the tube.

(*b*) *The sessile drop method*

This is another static method that is frequently used for mercury. Equation (7.14) can be used, though h is quite difficult to determine. The risk of the surface becoming contaminated is high. Unless empirical corrections are applied to equation (7.14), drops of very large diameter must be used. The angle of contact θ can be obtained from equation (7.15) once γ is known.

(*c*) *Jaeger's method*

This is a dynamic method involving the formation of a bubble at the end of a tube dipping into the liquid concerned. Fig. 7.21 shows a possible experimental arrangement. A tube of radius R dips into a liquid of density ρ so that the tip of the tube is a distance d below the surface. The excess pressure of the gas in the tube over the atmospheric pressure may be measured on the manometer containing a light oil of density σ. As the gas pressure in the tube is increased a bubble is produced at the end of the tube dipping into the liquid. The pressure in this bubble is a maximum when its radius of curvature is a minimum, that is, when it is equal to the radius R of the tube. When this size is reached, any increase in bubble size corresponds to a lower pressure

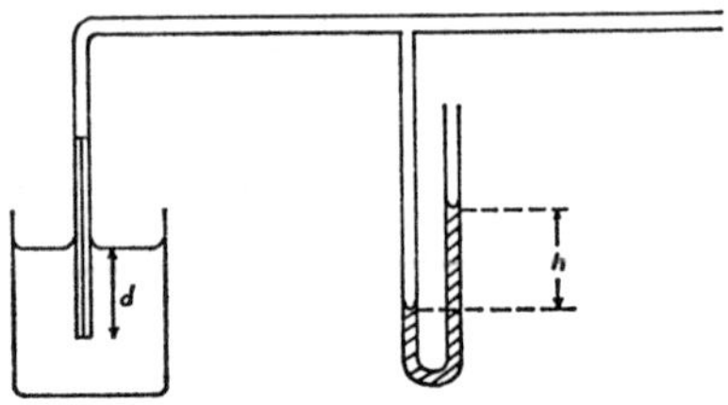

Figure 7.21

in the bubble for equilibrium, so the bubble rapidly becomes unstable and breaks away. Let h be the maximum difference in levels of the arms of the manometer. This occurs when the bubble is a hemisphere of radius R.

Then
$$\frac{2\gamma}{R} + d\rho g = h\sigma g$$

or,
$$\gamma = Rg(h\sigma - d\rho)/2 \tag{7.16}$$

This equation is, in fact, approximate, since the bubble is flattened by the variations in hydrostatic pressure over its surface, and it has a tendency to spread over the thickness of the tube. For this reason the method is most widely used for the comparison of free surface energies. It is very convenient for the investigation of the temperature variation of γ and has been widely used for measurements on molten metals. The bubble surface is always freshly formed.

(*d*) *Drop weight method*

This method utilizes the fact that when drops of a liquid are allowed to form slowly at the end of a capillary tube with a horizontal circular tip and then break away they are of a constant size. The break-away, in fact, often produces several drops; the main drop is followed by one or more much smaller satellite drops.

Harkins and Brown used dimensional analysis to show that the weight of the drop (including the satellites) that breaks away after forming slowly, mg, is related to the radius of the tip of the tube r, and the density of the liquid, ρ, by

$$\frac{mg}{\gamma r} = f\left(\frac{m}{\rho r^3}\right) \tag{7.17}$$

where f is an unknown function. They evaluated f for water and other liquids and presented the results graphically. Points for all liquids lie on a common curve. The essential part of the apparatus is easy to clean and it is not necessary to know the angle of contact. No matter how slowly the drop forms it breaks away quickly, so that the viscosity of

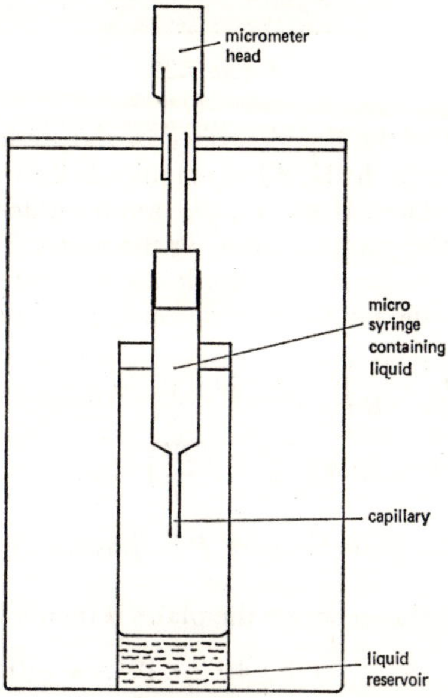

Figure 7.22

the liquid will have some effect on the size of the drop, but this is usually negligible. Since the drop forms slowly it is important to prevent evaporation and this is done by performing the measurement in an enclosed volume. A suitable experimental arrangement is shown in fig. 7.22. In this the liquid is slowly forced through a capillary tube attached to a micro-syringe as the result of a piston being moved by a micrometer screw. The movement of the screw between the break-away of consecutive drops enables the drop volume to be calculated. If the liquid wets the material of the capillary, r is the external radius of the capillary wall. On the other hand, if the liquid is non-wetting, r is the internal radius. This method is also suitable for measuring free interfacial energies. In this case the reservoir contains the liquid against which the free surface energy is being measured and the tip of the narrow tube is submerged in it.

7.11 Examples

(*a*) Two circular glass plates, each of 0·05 m radius, are separated by a water film $0{\cdot}01 \times 10^{-3}$ m thick. Find the minimum direct pull needed to separate them. The free surface energy of water at the temperature of the experiment is $73{\cdot}0 \times 10^{-3}$ J m^{-2} and the angle of contact of water with glass is zero.

A diametrical section of the glass plates and water film is shown in fig. 7.23.

Figure 7.23

The curvature of the liquid surface in the section shown gives rise to a pressure defect in the liquid of magnitude $2\gamma/r$ where $2r$ equals the separation of the plates. However, the liquid surface is also curved in a plane parallel to the planes containing the plates. This curvature gives rise to a pressure excess of magnitude $2\gamma/R$, where R is the radius of the plates. Hence, the net pressure defect in the liquid is of magnitude

$$p = 2\gamma\left(\frac{1}{r} - \frac{1}{R}\right) = 2 \times 73 \times 10^{-3}\left(\frac{10^3}{0{\cdot}005} - \frac{1}{0{\cdot}05}\right) \text{ N m}^{-2}$$

$$= 2 \times 73 \times 10^{-3}\left(\frac{10^6}{5} - \frac{10^2}{5}\right) \text{ N m}^{-2}$$

The effect of the finite value of R is thus seen to be very small and will be neglected.

The force needed to separate the plates is then $p \times$ (area of one plate)

$$= \frac{146 \times 10^{-3} \times 10^6 \times \pi \times (0{\cdot}05)^2}{5} \text{ N}$$

$$= 229 \text{ N}$$

(*b*) If the latent heat of vaporization of water is 2270×10^3 J kg^{-1}, its density $1{\cdot}0 \times 10^3$ kg m^{-3} and its free surface energy 73×10^{-3} J m^{-2}, calculate the minimum radius a drop can have if it is free to evapoarate but is thermally isolated.

Consider a drop, assumed spherical, of radius r. Let its radius decrease by an amount δr. The mass of liquid removed is $4\pi r^2 \delta r \rho$, where ρ is the liquid density, and the energy needed to do this is $4\pi r^2 \delta r \rho L$ where L is the latent heat of vaporization. This evaporation will occur spontaneously if the necessary energy is obtained from the decrease in free surface energy that results from the reduction of the surface area. The change in surface area during the process considered is $8\pi r\, \delta r$ corresponding to a change in free surface energy of $8\pi\gamma r\, \delta r$. Thus spontaneous evaporation will occur when

$$8\pi\gamma r \delta r = 4\pi r^2 \delta r \rho L$$

that is, when

$$\frac{2\gamma}{\rho L} = r$$

Thus, $$r = \frac{2\times73\times10^{-3}}{10^3\times2270\times10^3}\ \text{m}$$

which gives $$r = 64\times10^{-12}\ \text{m}$$

Hence r is of the same order of magnitude as the diameter of a molecule, and it is doubtful if γ has any real meaning under such circumstances.

Problems

1. A tube has a circular cross-section which varies from d_1 at one end to d_2 at the other ($d_1>d_2$). Plane soap films are formed across the ends of the tube and the internal pressure is slowly increased by admitting air through a side tube. What happens to the two soap films?

2. A vertical capillary tube of internal radius a stands vertically with its lower end dipping into a liquid of density ρ and free surface energy γ_1. The upper end of the tube is closed by a soap bubble of radius b, the free surface energy of the soap solution being γ_2. Derive an expression for the height of the liquid column, assuming that the liquid wets the material of the tube.

3. A capillary tube of internal diameter $0{\cdot}4\times10^{-3}$ m stands vertically in water of free surface energy 72×10^{-3} J m^{-2}. The angle of contact for water and the material of the tube is zero. The tube is slowly depressed vertically until a length of only 0·02 m is above the outside liquid level. Describe what happens.

4. A small quantity of water is placed in a glass tube that has a narrow conical bore. The angle of contact between glass and water is zero. The tube is held vertically and it is found that the water is retained in the mid portion of the tube. Which end is uppermost when the water is retained in this way?

5. A loop of fine thread is placed upon a plane soap film and the film inside the loop is then broken. Show that the loop will form a circle and derive an expression for the tension in the thread in terms of the free surface energy γ of the film and the radius of the loop r.

Further reading

Burdon, R. S., *Surface Tension and the Spreading of Liquids*, second edition, Cambridge Monographs on Physics, C.U.P., 1949.

Feather, N., *Mass, Length and Time*, Penguin Books Ltd, 1961, ch. 17.

Yarnold, G. D., *Contemporary Physics*, Volume 1, p. 267, 1960.

appendix: si units

The International System of Units is a coherent system based on six basic units: the metre, kilogram, second, ampere, degree Kelvin and candela. The units of the system are called SI units.

The SI unit of force is that force which, acting on a mass of 1 kilogram, produces an acceleration of 1 metre per second per second. The name newton is given to this force and its symbol is N.

A force of 1 N produces an acceleration of 1 m s^{-2} in a mass of 1 kg. Thus it produces an acceleration of 100 cm s^{-2} in a mass of 1000 g and hence is equal to a force of 10^5 dyn. Hence, 1 dyn $cm^{-2} = 10^{-5}$ N cm^{-2} $= 10^{-5} \times 10^4$ N $m^{-2} = 0{\cdot}1$ N m^{-2}.

When a force of 1 N moves a distance of 1 m in its own direction it performs the SI unit of work. A N m $= 10^5 \times 10^2$ dyn cm $= 10^7$ erg $=$ 1 J. Thus the SI unit of work is the joule.

In a similar manner the relationship between other SI units and the corresponding c.g.s. units can be found.

The units and names, where appropriate, of the quantities used in this book are given in the table below.

Quantity	*Name of Unit*	*Symbol of Unit*
mass	kilogram	kg
length	metre	m
time	second	s
area		m^2
volume		m^3
acceleration		m s^{-2}
angle	radian	
force	newton	N = kg m s^{-2}
pressure		N m^{-2}
stress		N m^{-2}
work, energy	joule	J = N m

Quantity	*Name of Unit*	*Symbol of Unit*
power	watt	$W = J\ s^{-1}$
torque		$N\ m$
free surface energy		$J\ m^{-2}$
dynamic viscosity		$N\ s\ m^{-2}$
kinematic viscosity		$m^2\ s^{-1}$

Conversion data

$1\ \text{dyn cm}^{-2} = 0{\cdot}1\ N\ m^{-2} = 0{\cdot}1\ J\ m^{-3}$

$1\ \text{torr} = 133\ N\ m^{-2}$

$1\ \text{erg} = 1\ \text{dyn cm} = 10^{-7}\ N\ m = 10^{-7} J$

$1\ cP = 10^{-3}\ N\ s\ m^{-2}$

$1\ cS = 10^{-6}\ m^2\ s^{-1}$

answers and hints

Chapter 2

1. Apply the following stresses to a cube of material with its faces normal to Ox, Oy, Oz respectively: σ along Ox, σ_1 along Oy, σ_1 along Oz. Using the theorem of superposition gives the following values for the strains: the strain in the x direction is $\frac{1}{E}[\sigma - 2\nu\sigma_1]$ that in the y direction is $\frac{1}{E}[\sigma_1 - \nu(\sigma+\sigma_1)]$ and that in the z direction $\frac{1}{E}[\sigma_1 - \nu(\sigma+\sigma_1)]$ where E is Young's modulus for the material and ν is Poisson's ratio. Choose σ_1 so that the strains in the y and z directions are zero. Then, $\sigma_1 = \nu\sigma/(1-\nu)$. The strain in the x direction is then $\frac{\sigma}{E}\left[1 - \frac{2\nu^2}{1-\nu}\right]$ and this is equal to σ/χ, where χ is the axial modulus of elasticity. Thus,

$$\frac{1}{\chi} = \frac{1}{E}\left[1 - \frac{2\nu^2}{1-\nu}\right] = \frac{1}{E}\frac{(1+\nu)(1-2\nu)}{(1-\nu)}$$

Combining this equation with equations (2.4) and (2.5) gives the required result.

2. $8{\cdot}8 \times 10^3$ N m. Use equation (2.8).

3. (a) $2{\cdot}45 \times 10^{-3}$ J (b) $4{\cdot}91 \times 10^{-3}$ J.

4. $1{\cdot}51 \times 10^{-3}$ J. Calculate the stored energy when an initially unloaded wire has a load of 1·5 kg applied. Repeat the calculation for an initially unloaded wire that has a load of 2·5 kg applied. The difference between these two values gives the change in stored energy when the load carried by the wire is increased from 1·5 kg to 2·5 kg.

5. Let y be the displacement of a molecule measured from its equilibrium position. If F is the intermolecular force and U the potential energy of the interaction

$$F = -dU/dy$$

Also, if α is the force constant, for small displacements $F = -\alpha y$. Combining these two expressions for F and integrating gives the variation of U with y.

6. $3{\cdot}2\times10^{12}$ Hz. Use the expression for the frequency given in section 2.1.

Chapter 3

1. (*a*) 22×10^{-3} J, (*b*) $5{\cdot}1\times10^{-6}$ J, (*c*) $5{\cdot}99\times10^{-3}$ m.

2. Construct a cube of side y in the material so that the dislocation line is parallel to an edge of the cube and the glide plane is parallel to a face of the cube. Let the dislocation move, on its glide plane, from one cube face to the opposite one. If F is the force acting on unit length of the dislocation line, then, since the dislocation of length y moves a distance y, the work done by the force is Fy^2. The movement of the dislocation produces a displacement of the cube surface above the glide plane and parallel to it of b relative to the opposite face below it. Hence the work done is also the shearing force τy^2 times the displacement, that is $\tau y^2 b$. Hence $Fy^2 = \tau y^2 b$ or $F = \tau b$.

4. The load F acting on a sample at any instant is equal to σa where σ is the true stress and a is the area of cross-section. Necking occurs when the load reaches a maximum value, that is, when $dF = d(\sigma a) = 0$.

Hence $$\frac{d\sigma}{\sigma} = -\frac{da}{a}$$

In plastic deformation the volume V of the sample remains constant. If l is the length of the sample

$$V = al = \text{constant}$$

so that $$dV = d(al) = 0$$

Hence $$\frac{da}{a} = -\frac{dl}{l}$$

and $$\frac{d\sigma}{\sigma} = \frac{dl}{l}$$

The tensile strain is $\epsilon = \dfrac{l-l_o}{l_o} = \dfrac{l}{l_o} - 1$

where l_o is the unstretched length of the specimen.

Thus, $$\frac{dl}{l} = \frac{dl}{l_o}\times\frac{l_o}{l} = \frac{d\epsilon}{1+\epsilon}$$

giving $$\frac{d\sigma}{\sigma} = \frac{d\epsilon}{1+\epsilon} \text{ or } \frac{d\sigma}{d\epsilon} = \frac{\sigma}{1+\epsilon}$$

which is the required result. Notice that this construction assumes that the purely elastic deformation is negligible.

5. Use Considère's construction. If F_m is the maximum load in a tensile test and a_m is the corresponding area of cross-section, the value of the true stress is $\sigma_m = F_m/a_m$. The ultimate tensile strength is $\sigma_u = F_m/a_o$ where a_o is the initial area of cross-section, so that

$$\sigma_u = \frac{F_m}{a_m} \times \frac{a_m}{a_o} = \sigma_m \times \frac{l_o}{l_m} = \frac{\sigma_m}{1+\epsilon_m}$$

where ϵ_m is the strain when necking starts.

Hence,
$$\frac{\sigma_u}{\sigma_m} = \frac{1}{1+\epsilon_m}$$

$$\sigma_m = 30\times10^6 \text{ kg m}^{-2}; \; \epsilon_m = 1; \; \sigma_u = 15\times10^6 \text{ kg m}^{-2}.$$

6. Let a glide packet of small thickness t and length l be bent to a mean radius of curvature r. The difference in length between the curved surfaces of the glide packet and its neutral plane of bending is $\pm \dfrac{tl}{2r}$ so that the misfit between two adjacent glide packets is tl/r. If this misfit is produced by n dislocations with Burgers vector $\boldsymbol{b}$ (parallel to the neutral plane), $nb = tl/r$ so that $n = tl/rb$. Thus the density of dislocations needed is $\dfrac{tl}{rb\,tl} = \dfrac{1}{rb}$

Chapter 5

1. 105 N m^{-2}.

2. (*a*) $1{\cdot}42\times10^{-3}$ N s m^{-2}, (*b*) $1{\cdot}30\times10^{-3}$ N s m^{-2}.

3. $(4\eta lA\ln 2)/\pi r^4\rho g$. Use the method of the example in 5.12 (*a*). If y_1 and y_2 are the heights of the liquid columns measured from the axis of the horizontal tube, initially y_1 is $3h$ and finally it is $5h/2$. At any instant of time t when $y_1 = y$ the value of y_2 is $4h-y$, so that $y_1-y_2 = 2y-4h$ and the pressure difference giving rise to the liquid flow is $2(y-2h)\rho g$. Then,

$$-A\frac{dy}{dt} = \frac{2(y-2h)\rho g\pi r^4}{8\eta l}$$

so that
$$-\int_{3h}^{5h/2} \frac{dy}{y-2h} = \frac{\rho g\pi r^4}{4\eta lA}\int_0^T dt$$

$$T = \frac{4\eta lA}{\pi r^4\rho g}\ln\left(\frac{3h-2h}{5h/2-2h}\right)$$

4. $1{\cdot}47$ N s m^{-2}. Use equation (5.7), obtaining the value of v_t for an infinite medium as indicated at the end of section 5.10.

5. $(\pi\eta\omega r^4)/2d$. Consider an element of A a distance b from the common axis, of thickness db and subtending a small angle $d\varphi$ at the axis. The speed of this element is $v = \omega b$, so that the velocity gradient in the liquid between this element and the corresponding one in B is $\omega b/d$. Assuming laminar liquid flow, the force that must be applied to this element to produce the rotation is $dF = \eta \times \text{area} \times \text{velocity gradient}$, that is,

$$\eta \times b d\varphi db \times \frac{\omega b}{d}$$

This force has a moment about the axis of

$$dC = \frac{\eta b^2 d\varphi db\omega}{d} \times b$$

so that the total couple needed to rotate the plate is

$$C = \int_0^{2\pi} d\varphi \int_0^r \frac{\eta\omega b^3}{d}\, db = \frac{\pi\eta\omega r^4}{2d}$$

6. Consider a cylindrical surface in the liquid of radius r and coaxial with the cylinders of radii a and b. If the angular velocity of the liquid at a distance r from the common axis is ω, the rate of shear is $r\dfrac{d\omega}{dr}$ and hence the force acting per unit area is $\eta\, r\dfrac{d\omega}{dr}$, assuming laminar liquid flow. If the inner cylinder is rotating in a clockwise direction, the liquid outside the cylindrical surface of radius r exerts a torque $2\pi\eta l r^3 \dfrac{d\omega}{dr}$ on that surface in a clockwise direction. If T is the clockwise couple needed to keep the inner cylinder rotating with angular velocity ω_0, then, since the liquid between this cylinder and that of radius r drawn in the liquid has zero angular acceleration,

$$T + 2\pi\eta l r^3 \frac{d\omega}{dr} = 0$$

and hence

$$-T\int_0^b \frac{dr}{r^3} = 2\pi\eta l \int_{\omega_0}^0 d\omega$$

so that

$$T\left[\frac{1}{a^2} - \frac{1}{b^2}\right] = 4\pi\eta l\omega_0$$

Chapter 6

1. 94×10^{-10} m. Use equations (6.1) and (6.5), remembering that the mean velocity of the molecules is $\sqrt{\dfrac{8}{3\pi}}$ times the root mean square velocity.

2. Use equation (16.7). $V_1 = \frac{l\pi a_1^2}{t}$, $p_1 = (P+x)\rho g$, $p_2 = P\rho g$.

3. $1{\cdot}77 \times 10^{-5}$ N s m^{-2}. Use equation (5.7), assuming that the raindrops remain spherical.

4. $2{\cdot}2 \times 10^{-10}$ m. Use equations (6·1), (6·4) and (6·5), together with $\bar{c} = \sqrt{\frac{8}{3\pi}}\, c_m$. In questions 1 and 4, more refined equations give answers differing slightly from those obtained here.

Chapter 7

1. The radii of curvature must remain equal for the two soap films as long as their surfaces remain parts of spheres. When the bubble at the end of larger bore becomes hemispherical its excess pressure is a maximum, so that a further increase in its size needs a drop in pressure for equilibrium. If air continues to be supplied, instability of the larger bubble is likely.

2. $h = \frac{1}{\rho g}\left(\frac{2\gamma_1}{a} - \frac{4\gamma_2}{b}\right)$.

Let atmospheric pressure be P and the pressure inside the bubble be p. Then,

$$p - \frac{4\gamma_2}{b} = P$$

and

$$p - \frac{2\gamma_1}{a} + h\rho g = P$$

Eliminating p and P from these equations gives the result.

3. The meniscus is at the top of the tube with a radius of curvature $0{\cdot}73 \times 10^{-3}$ m.

4. The narrow end is uppermost. Let the length of the column of water be h. At the upper end of this column let the radius of the tube be r_1 and at the lower end be r_2. In equilibrium the pressure acting downwards on a horizontal plane in the liquid must be equal to the upward pressure on the same plane. Let such a plane be a distance h_1 below the upper surface of the liquid. Then, if P is the atmospheric pressure,

$$P - \frac{2\gamma}{r_1} + h_1\rho g = P - \frac{2\gamma}{r_2} - (h - h_1)\rho g$$

or,

$$h\rho g = 2\gamma\left(\frac{1}{r_1} - \frac{1}{r_2}\right)$$

where γ is the free surface energy of water and the angle of contact with glass is assumed to be zero. h must be positive and hence r_2 is greater than r_1.

5. The soap film will contract so that its area is a minimum, that is, so that the area inside the loop of thread is a maximum. For a given periphery, a circle has the largest area.

Apply the principle of virtual work to the system. If T is the tension in the thread, then, for an infinitesimal change in the length of the thread, $-T\times$(change in length of thread)$+\gamma\times$(change in surface area of soap film) $= 0$,

so that $$-T\times 2\pi dr+\gamma\times 2\times 2\pi r dr = 0$$

or, $$T = 2\gamma r$$

Remember that work must be done on the thread to increase its length, whereas work is done by the soap film when its area is reduced and a film has two surfaces.

index

index

Only one reference is given under a particular heading per section. The numbers refer to pages.

CRANFIELD INSTITUTE OF TECHNOLOGY
LIBRARY

FIELD INSTITUTE OF TECHNOLOGY
LIBRARY